Handbook of Circuits and Systems

Handbook of Circuits and Systems

Jasper Harrison

NY RESEARCH PRESS
New York

Published by NY Research Press
118-35 Queens Blvd., Suite 400,
Forest Hills, NY 11375, USA
www.nyresearchpress.com

Handbook of Circuits and Systems
Jasper Harrison

International Standard Book Number: 978-1-63238-849-0 (Hardback)

Cataloging-in-Publication Data

Handbook of circuits and systems / Jasper Harrison.
p. cm.
Includes bibliographical references and index.
ISBN 978-1-63238-849-0
1. Electronic circuits. 2. Electronic systems. 3. Electronics. I. Harrison, Jasper.
TK7867 .H36 2022
621.381 5--dc23

Contents

Preface

A system of electrical components and conductors through which electrical current flows is known as a circuit, or an electronic circuit. Some of the different components which can be a part of circuits are transistors, inductors, capacitors, resistors and diodes. There are a variety of operations which can be performed through the use of circuits such as amplifying signals and moving data. Circuits can either be constructed of individual connected wires or the interconnections can be made upon laminated substrates using photolithographic techniques. Some of the different types of circuits which are used to test new designs are perfboards, breadboards and stripboards. These circuits are easy to redesign, making it easy for designers to make changes quickly. This book discusses the fundamentals as well as modern approaches of circuits and their systems. It will serve as a valuable source of reference for graduate and post graduate students. The book is appropriate for students seeking detailed information in this area as well as for experts.

A short introduction to every chapter is written below to provide an overview of the content of the book:

Chapter 1 - The closed pathway which functions as a medium for the flow of electrons is known as a circuit. They are broadly divided into two categories, namely, electrical circuits and electronic circuits. This is an introductory chapter which will introduce briefly all the significant aspects of circuits.; **Chapter 2** - There are various components which can be present in a circuit. These are divided into two categories, passive elements and active elements. Some of the passive elements are resistors, capacitors, inductors and diodes. This chapter discusses in detail these basic elements of circuits, along with active sources such as voltage sources and current sources.; **Chapter 3** - There are a number of theorems and laws pertaining to circuits such as Ohm's law, Norton's theorem and Superposition Theorem. Ohm's law provides the relationship between current, resistance and voltage. The diverse applications of these laws and theorems in different scenarios have been thoroughly discussed in this chapter.; **Chapter 4** - A physical interconnection of different components which is involved in gathering various amounts of information together is known as an electronic system. A system which manages, directs and regulates the behavior of other devices is known as a control system. The chapter closely examines these key concepts of electronic and control systems to provide an extensive understanding of the subject.; **Chapter 5** - System function or transfer function is used to show the relationship between the input signal and the output signal for all the possible input values. Some of the methods for obtaining a transfer function are block diagram method and signal flow graphs. This chapter discusses in detail these theories and methodologies related to system theory and functions.

I extend my sincere thanks to the publisher for considering me worthy of this task. Finally, I thank my family for being a source of support and help.

Jasper Harrison

1

Understanding Circuits

The closed pathway which functions as a medium for the flow of electrons is known as a circuit. They are broadly divided into two categories, namely, electrical circuits and electronic circuits. This is an introductory chapter which will introduce briefly all the significant aspects of circuits.

The term circuit refers to a physical or conceptual loop, often with an emphasis on resolving or closing that loop to accommodate various operations associated with electrical systems or other technologies. In IT, the term is often used when discussing the physical manufacturing of devices or pieces of hardware.

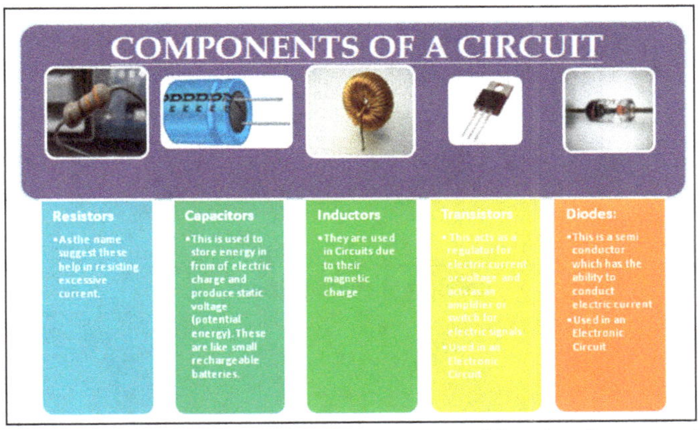

The components of a circuit are arranged in many ways, two of the simplest are – Series and Parallel.

Electronic Circuit

An electronic circuit consists of various electronic components like resistors, capacitor, diodes and transistors connected by a wire, through which current flows in the circuit. The electronic circuit design is generally designed on a breadboard first (prototyping) that helps the designer for modification and enhancement of the circuit. These electronic circuits are used in computations, data transfer and signal amplifications.

An Electronic Circuit Approach on Breadboard and PCB

Basics of an Electronics Circuit Design Process

An every elementary electronic device constructed as a single unit. Before the invention of digital circuits (ICs), all individual transistors, diodes, resistors, capacitors, and inductors were discrete in nature. Any circuit or a system can produce the preferred output based on its input.

Analog Circuit

Analogue electronic circuit designs are those in which current or voltage varies with time to correspond to the information being represented. Diodes, capacitors, resistors, transistors and wires are the major components of an analog circuit. In analog circuits, electrical signals take the continuous value, and these circuits are represented in schematic diagrams, where wires are represented by lines and each component is represented by unique symbols. Every analog circuitry has series or parallel or both circuits.

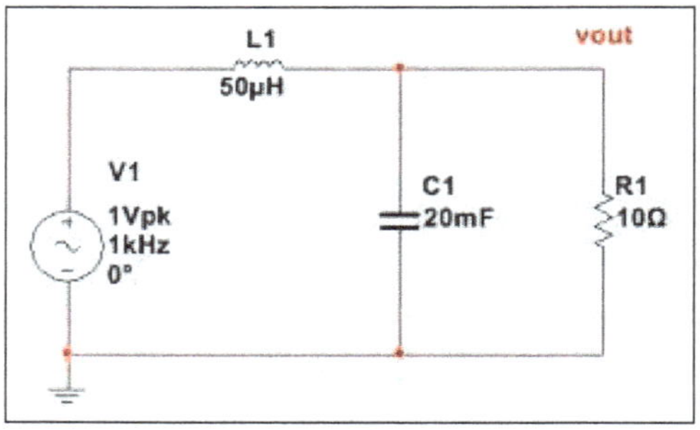

A simple analog circuit

Digital Circuits

Digital electronic circuit design takes the electrical signals in the form of discrete values. The data are represented in the form of zeros and ones. Digital circuits extensively use transistors, interconnected to give create logic gates that provide the function of Boolean logic. Transistors are interconnected to provide the positive feedback as used in latches and flip-flops. Therefore digital circuits can provide both logic and memory, enabling them to perform computations.

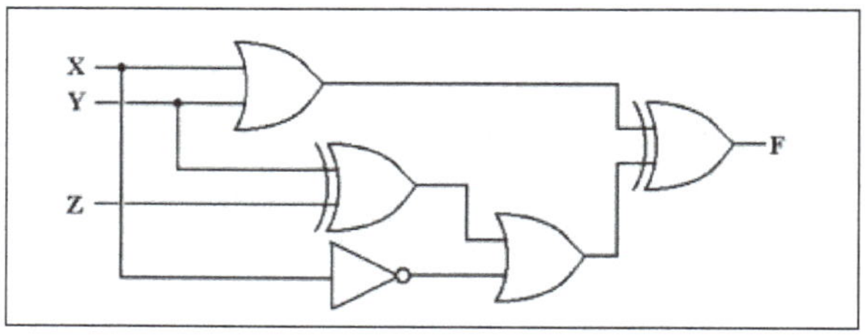

Digital circuit using flip-flops

Digital circuit is used to create general purpose computing chips like microprocessors and application specific integrated circuits.

Difference between Analog Circuit and Digital Circuit

Analog and digital circuits are used to transmit and process the information like sound, light from an environment to generate continuous variable signals. Analog circuits can route the signals directly, whereas digital circuits change the analog signals by evaluating them at regular intervals and giving out the resulting values. To get the outputs, analog circuits can directly give the signals while a digital circuit has to change the information back to an analog signal. The working of an analog circuits can be done with a normal waveform for converting, storing and amplifying the signals, while a digital circuit alters the wave forms into pulse signals.

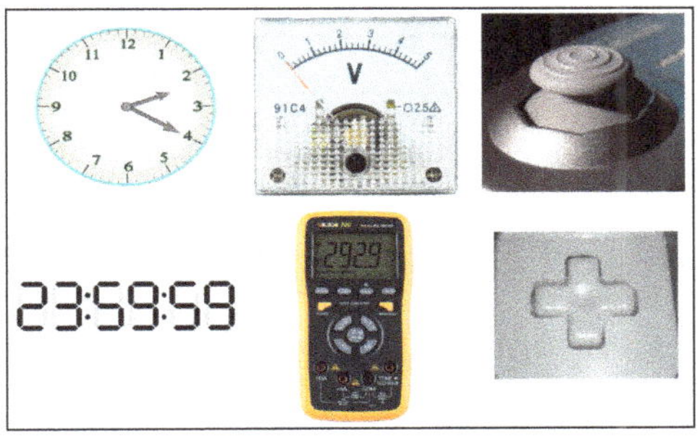

Analog vs digital

The main difference between analog circuit and digital circuit mainly includes what is an analog circuit, digital circuit and the differences between them.

Analog Circuit

The Analog electronic circuit includes an analog signal with any continuously changeable signal. While working on an analog signal, an analog circuit alters the signal in some manner. Analog circuit can be used to convert the original signal into some other format such as a digital signal. Analog circuits may also modify signals in inadvertent ways like adding noise or distortion. Analog circuits are classified into two types, namely active analog circuits and passive analog circuits. An analog circuit uses an electrical power source to get the goals of a designer while Passive circuits use no external electrical power.

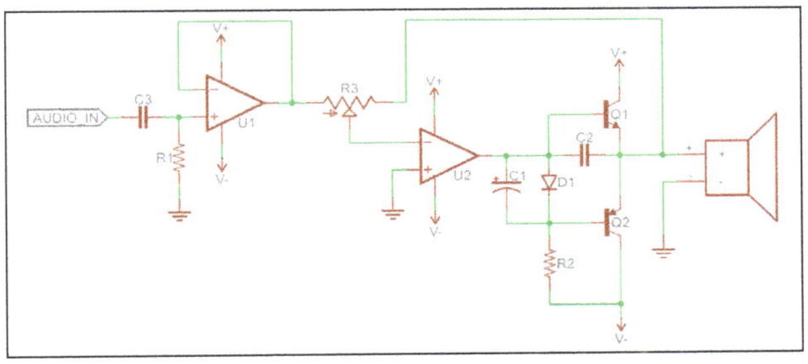

Analog circuit

Digital Circuit

A digital circuit is a circuit where the signal should be one of two discrete levels. Each level is interpreted as one of two different states (for instance, 0 or 1). These circuits built with transistors to make logic gates in order to execute Boolean logic operation. This logic is the base of digital electronics & computer processing. Digital circuits are less vulnerable to degradation in excellence than analog circuits. It is also simpler to execute error detection and rectification with digital signals. To make the routine process of designing these circuits, designers use EDA (electronic design automation) tools, a kind of software that develops the logic in a digital circuit.

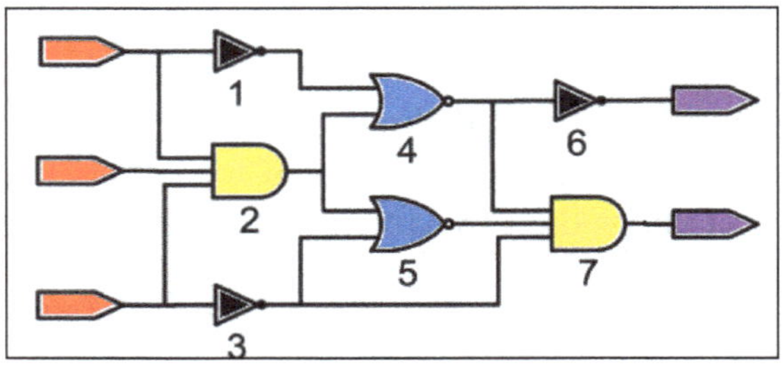

Digital circuit

Working of Analog Circuit and Digital Circuits

The working of analog circuits can be done with normal waveforms, changing them. For ex: a microphone is used in an analog circuit which converts the sound waves into analog electric waves. For example, in an analog circuit a microphone converts sound waves into similar or analog electric waves. These signals can be stored on the strip, improved in an analog amplifier and transformed back to related sound waves by a speaker.

A digital circuit is used to alter the waves into pulse waves. It measures a waveform thousand of times every second and stores the data in binary form. For instance, after 12msecs, a signal might be 2.4 volts high & after 14msecs, it might be at 2.6 volts. This circuit changes the volts and times into binary data, and the waves become a series of 1's and 0's. When the circuit has to generate sound from a speaker, it produces an o/p signal which is at 2.4 V after 12msec and at 2.6 V after 14msec, similar to the original wave.

O/P Quality

Analog circuit delivers an analog of the ordinary wave form and can generate very high o/p quality. Digital circuits generate an estimate of the natural waveform, so their quality of o/p is incomplete by the number of curve size they carry out.

Efficiency of a Circuit

An efficiency of a circuit mainly depends on how rapidly it can produce results and how much power it utilizes. These circuits have to wait for a wave to complete a cycle before they can carry out their tasks. And they require sufficient power to generate the peak of the wave. The digital circuit's speed is incomplete only by the speed of the gears that make up the circuit and not by the indications they are processing. To work with tiny pulses needs moderately little power. In most of the applications, analog circuits are slower & require more power than digital circuits.

Precision and Reproducibility

The act of analog circuits depends on their design and their gears to guarantee the waveform leftovers true to the original. They are liable to design faults, modifies from aging parts and outside powers such as electrical noise. Digital circuits just have to maintain path of their pulses. Even if numerous pulses get vanished, it influences only a few of the thousands of sizes.

Main Differences

The main differences between analog circuit and digital circuits mainly includes the following:

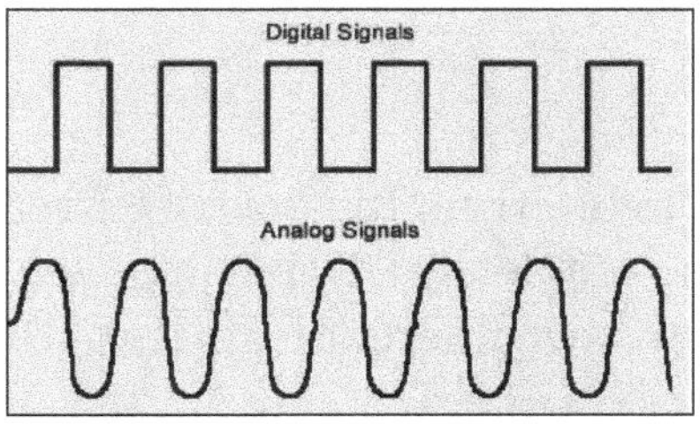

Differences between analog and digital

- Analog circuits operate on analog signals commonly known as continuous valued signals.
- Digital circuits function on signals that exist simply at 2 levels, i.e. Zeros and ones.
- The designing of an analog circuit is difficult since every component has to be positioned by hand for designing the circuits.
- Digital circuits are very simple to design since the technique of an automation can be useful at a variety of levels of circuit design.
- No change of i/p signals is necessary before processing, the circuit straightly executes different logical operations and generates an analog o/p.
- In digital circuits, the i/p signals change from analog to digital (A/D) form before it is processed, that is the digital circuit is accomplished by processing digital signals only, and generates o/p which is again changed back from digital to analog signals (D/A) so that the o/p gives relevant results that can be understood by individuals.
- Analog circuits are typically routine made and they don't have flexibility.
- Digital circuits have a high degree of elasticity.

Schematic Circuit Diagrams

A schematic circuit diagram is the representation of components and interconnections in a circuit using standardised symbols without using the actual image of the component. Circuit diagrams are used for design, construction and maintenance of the electrical and electronic equipment.

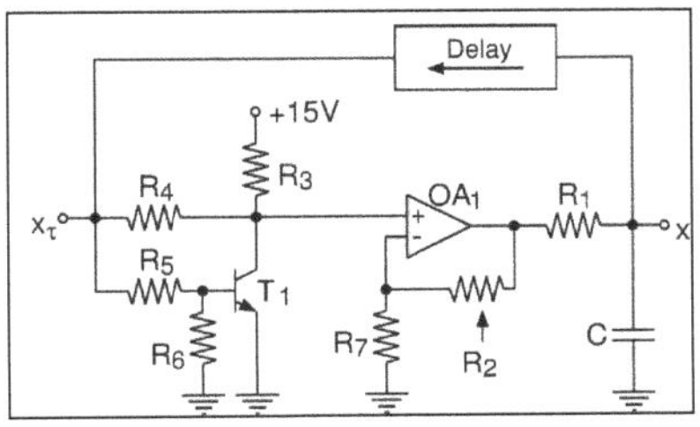

Schematic circuit diagrams

Though it is not standardised, the schematic diagrams are organised on a page from left to right and top to bottom. Like, in signalling circuitry the antenna is to the left and speaker to the right. Similarly, positive power supply at the top of the page, with ground and negative supply at the bottom. Relay logic line diagrams also use standardised methods to represent schematic diagrams. A vertical power supply rail at the left and another on the right with the components strung between them representing a ladder. Hence, it is also called as ladder logic diagram.

Electronic Switch Circuit

A switch is an electrical device used to interrupt the flow of current in the circuit. These are essentially binary devices which are either completely ON or completely OFF. Besides ON/OFF switches controls the work of a circuit and activates different features of the circuit.

Switches are the mechanical devices with two or more terminals that are connected to the metal contacts. When the contacts are together, the switch is closed. Thus the current flows and switch is ON. When the contact is apart, the switch is open and no current flows.

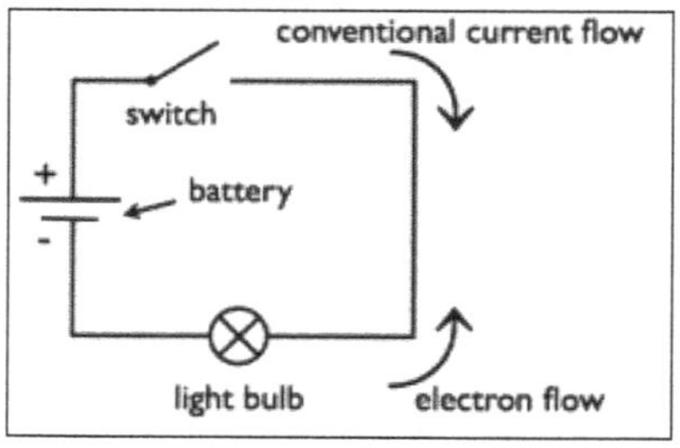

Electronic switch circuit

The above circuitry shows how the switch is used to control the current flow in the bulb. The below are the various switches used in the electronic circuits.

Toggle Switch

The toggle switch is actuated by a lever angled in one or more positions. The lever flips up or down to close or open the contact. The light switches used in the household are the example of a toggle switch.

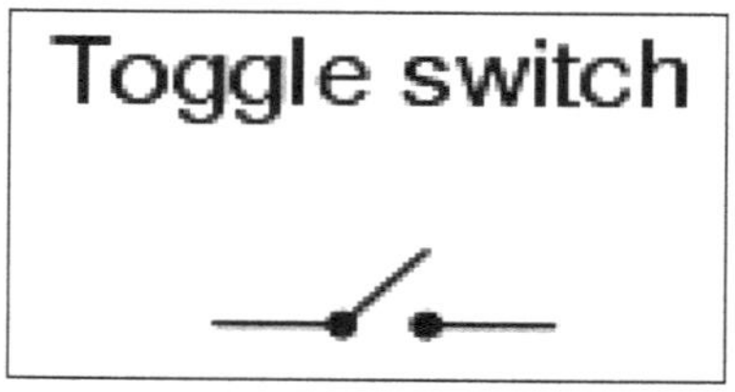

Toggle switch

Push Button Switch

Push button switch is a two-position device actuated with a button to open and close the contacts. Each time you push the button the contact alternates between the open and close.

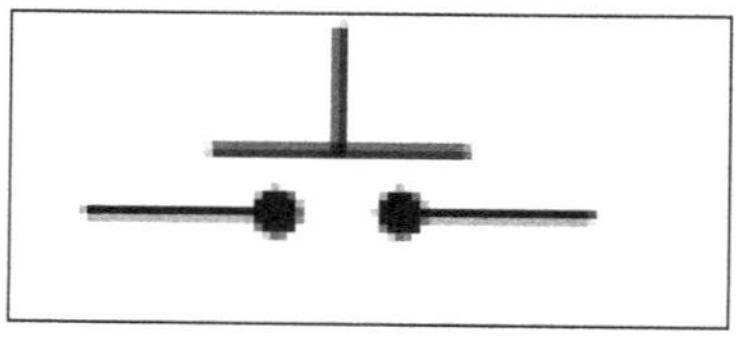

Push button switch

Selector Switch

Selector switches are actuated with a rotary knob or a lever to select one or two positions. Selector switch can either rest in any of their positions like the toggle switch.

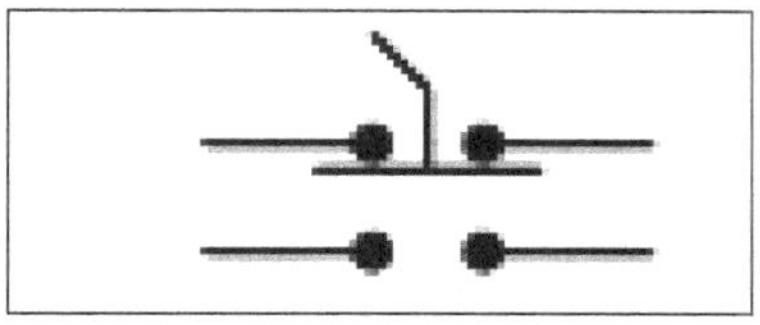

Selector switch

Joystick

A joystick switch is triggered by a lever free to move in more than one axis of motion. The circle and dot notation on the switch symbol indicate the direction of joystick lever motion that is required to trigger the contact. Joystick hand switches are used to control crane, robot and in games.

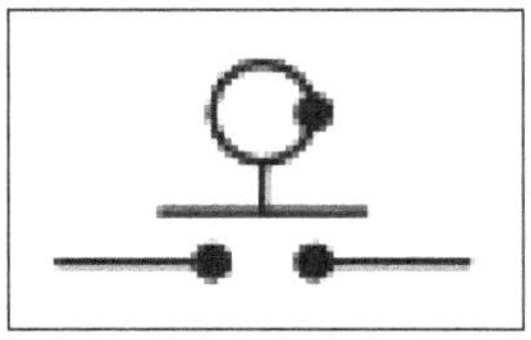

Joystick

Liquid Level Switch

A floating object is used to activate the switch mechanism when the liquid level rises to a fixed point. When the liquid level reaches a point, the floating object closes the circuit. This closed circuit conducts, making it perform the specific task.

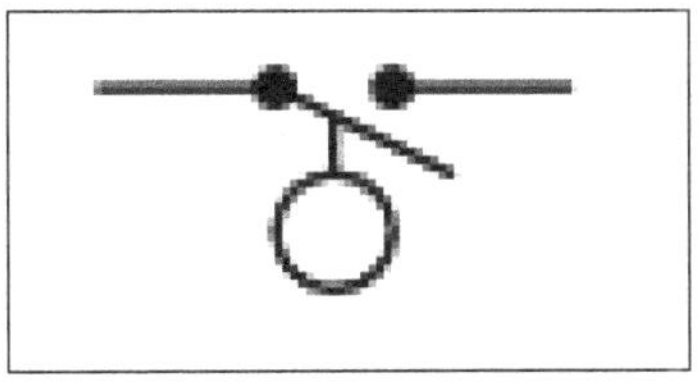

Liquid level switch

Lever actuator limit switch, pressure switch, proximity switch, speed switch and the nuclear level switch are various other switches used in electronic circuits.

Electronic Circuit Design

Electronic circuit design consists of analysis and synthesis of electronic circuits. While designing an analog circuit or digital circuit the designer should be able to predict the voltage and current at every node in the circuit. All linear circuits and simple non-linear circuits can be analysed by hand using mathematical computations. While software is used to analyse the complex circuits.

Electronic circuit design simulation software allows the developer to design circuits more effectively and accurately, further reducing the time, cost and risk involved in developing the circuit prototypes.

Circuit Board Simulator

The electronic circuit simulator uses mathematical models to replicate the behaviour of an actual electronic circuit. Simulation software allows for modelling of circuit operation and is an invaluable analysis tool. Because of the limitation of the breadboard and expensive tools like photo masks for integrated circuits, most of the IC design relies on simulation.

Though circuit board simulators make an ease of developing a large circuit they certain complexities in the simulation process. Process variations occur when a design is fabricated, but the circuit simulators do not consider these variations. Though the variations are small they affect the output significantly.

Electronic Circuit Symbols

In electronic circuits, there are many electronic symbols that are used to represent or identify a basic electronic or electrical device. They are mostly used to draw a circuit diagram and are standardized internationally by the IEEE standard and the British Standard. NO changes can be brought by the user on any electronic symbol, but the user is free to bring any changes in the architectural drawings like power source and lighting.

Electronic Symbols

The symbols for different electronic devices are shown below. Apart from the circuit symbols, each device is also designated a short name. Though these names are not approved as standard notations, they are commonly used by most people. These designations are also given in the list.

Wire Symbols

- Wire: Used to connect one component to another.

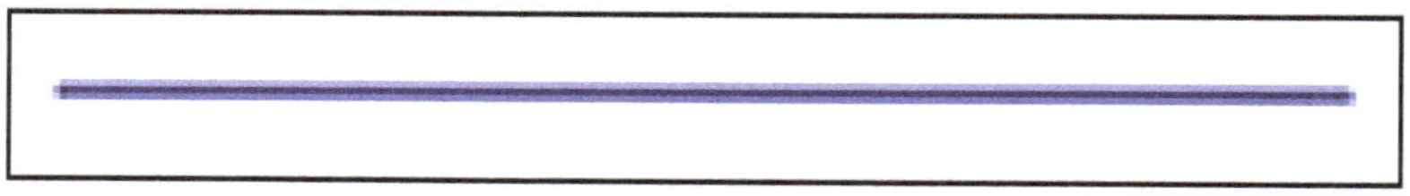

Wire circuit symbol

- Wires Joined: One device may be connected to another through wires. This is represented by drawing on the point where they are shorted.

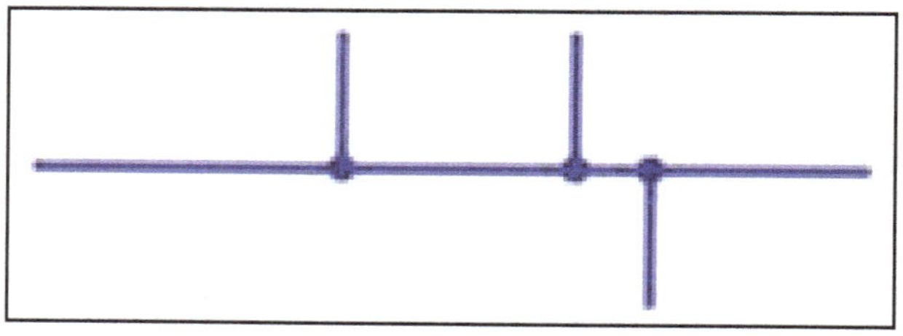

Wires joined circuit symbol

- Un-joined Wires: When circuits are drawn some wires may not touch others. This can only be shown by bridging them or by drawing them without blobs.

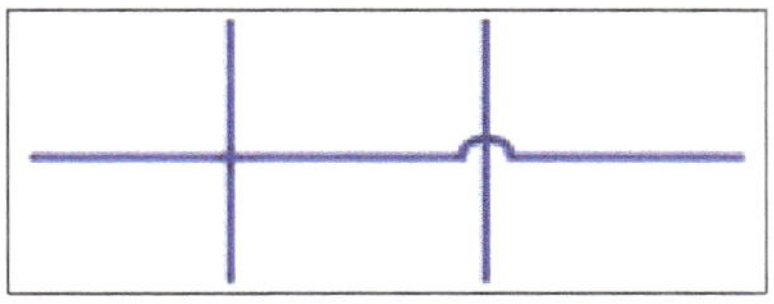

Wires not joined circuit symbol

Power Supply Symbols

- Cell: Used to provide a supply for a circuit.

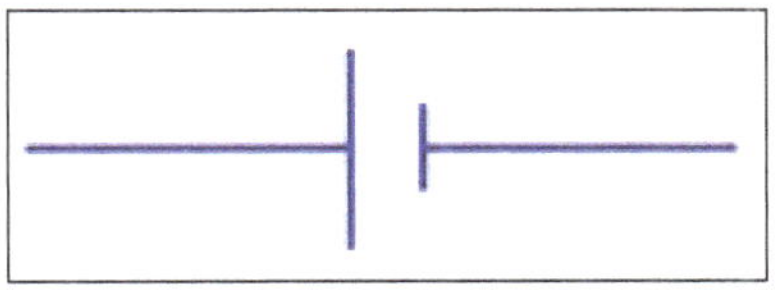

Cell circuit symbol

- Battery: A battery has more than a cell and is used for the same purpose. The smaller terminal is negative and the larger one is positive.

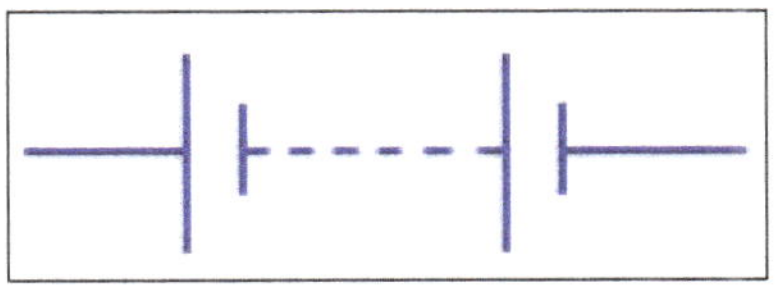

Battery circuit symbol

- DC Supply: Used as a DC power supply, that is, the current will always flow in one direction.

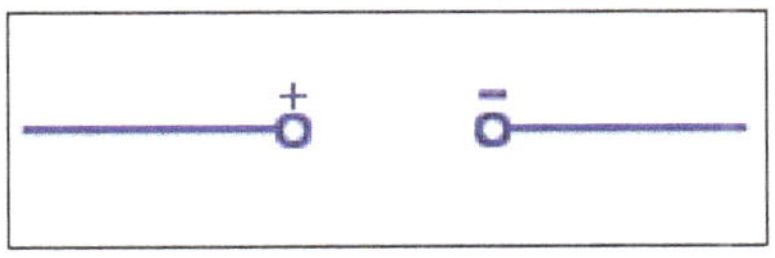

DC supply circuit symbol

- AC Supply: Used as AC power supply, that is, the current will keep alternating directions.

AC supply circuit symbol

- Fuse: Used in circuits where a probability of excessive current flows. The fuse will break the circuit if excessive current flows and saves the other devices from damage.

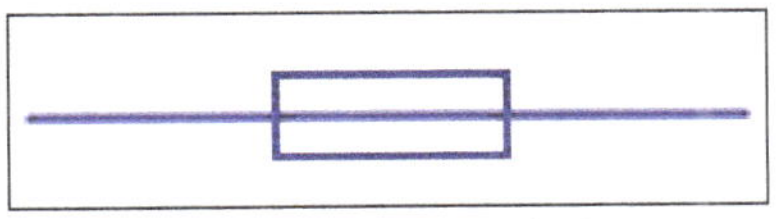

Fuse circuit symbol

- Transformer: Used as an ac power supply. Consists of two coils, the primary and secondary that are linked together through an iron core. There is no physical connection between the two coils. The principle of mutual inductance is used to obtain power.

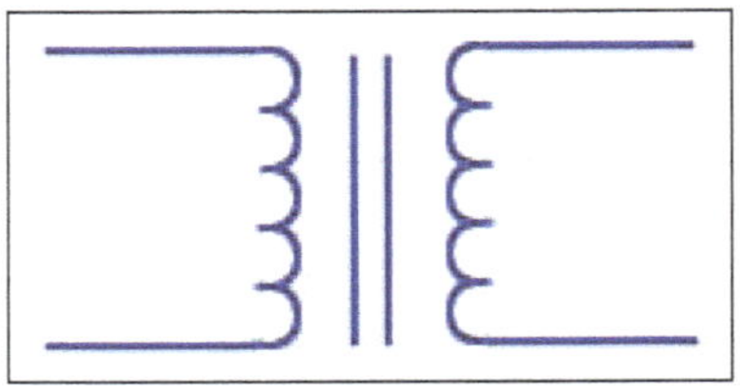

Transformer circuit symbol

- Earth/Ground: Used in electronic circuits to represent the 0 volts of the power supply. It can also be defined as the real earth, when it is applied in radio circuits and power circuits.

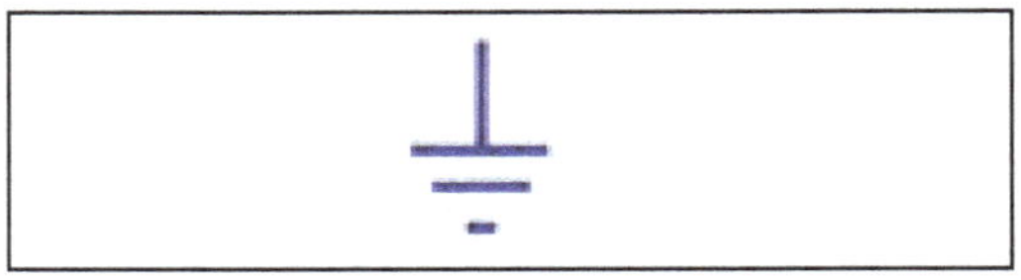

Earth circuit symbol

- Resistor Symbols Resistor: A resistor is used to restrict the amount of current flow through a device.

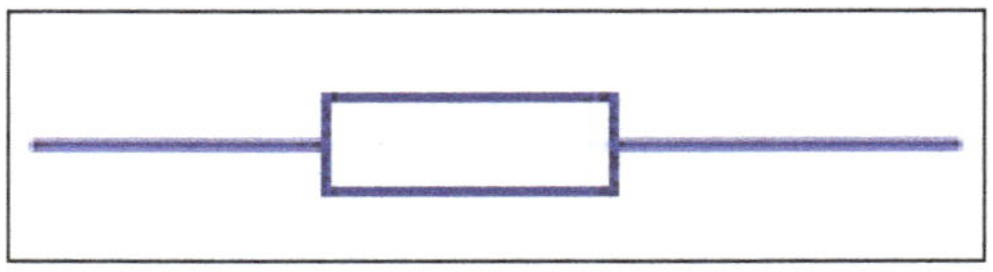

Resistor Circuit Symbol

- Rheostat: A rheostat is used to control the current flow with two contacts. Applicable in controlling lamp brightness, capacitor charge rate, etc.

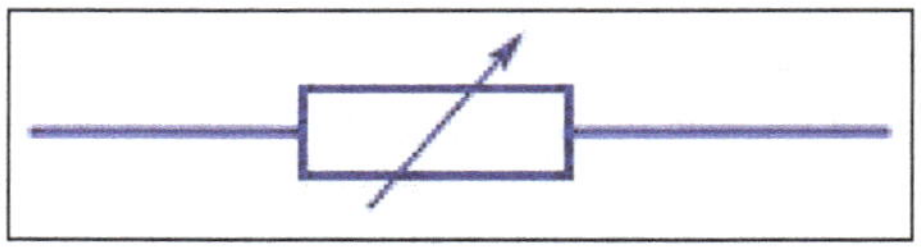

Rheostat circuit symbol

- Potentiometer: A potentiometer is used to control the voltage flow and has three contacts. Have applications in changing a mechanical angle change to an electrical parameter.

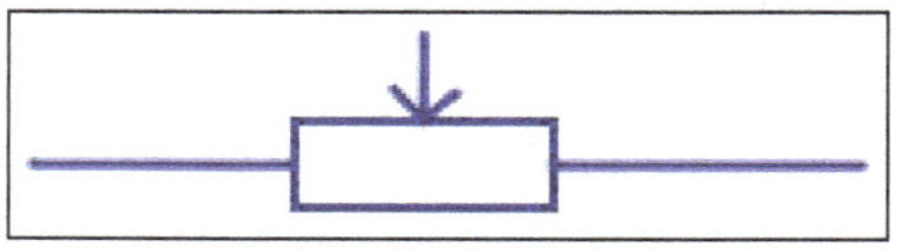

Potentiometer circuit symbol

- Preset: Presets are low cost variable resistors that are used to control the charge flow with the help of a screw driver. Applications where the resistance is determined only at the end of the circuit design.

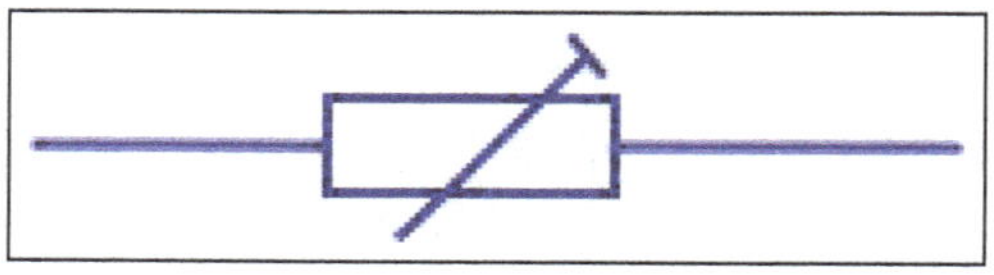

Preset circuit symbol

Capacitor Symbols

1. Capacitor: Capacitor is a device that is used to store electrical energy. It consists of two metals plates that are separated by a dielectric. It is applicable as a filter, that is, to block DC signals and allow AC signals.

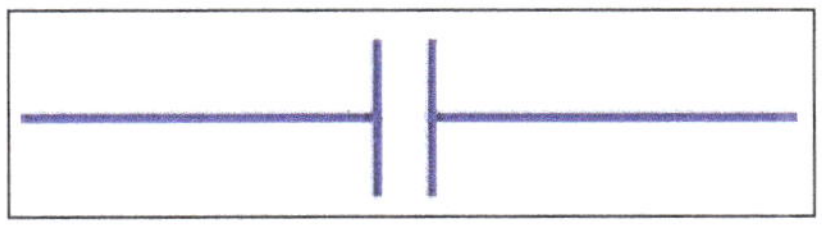

Capacitor circuit symbol

2. Capacitor: Capacitor can be used in a timer circuit by adding a resistor.

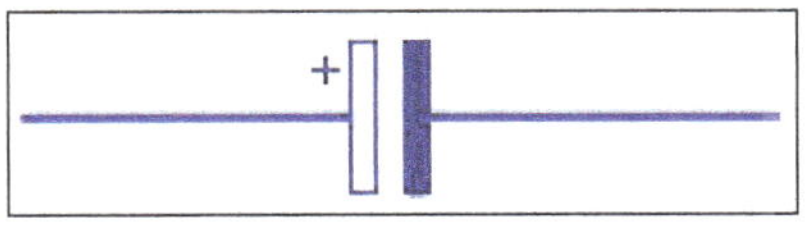

Capacitor-polarised circuit symbol

3. Variable Capacitor: Used to vary the capacitance by turning the knob. A type of variable capacitor is the trimmer capacitor that is small in size. The notations are all the same

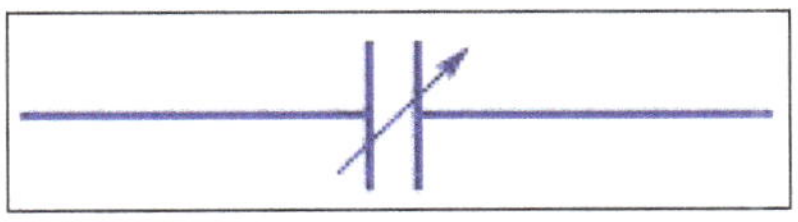

Variable capacitor circuit symbol

Diode Symbols

1. Diode: A diode is used to allow electric current to flow in only one direction.

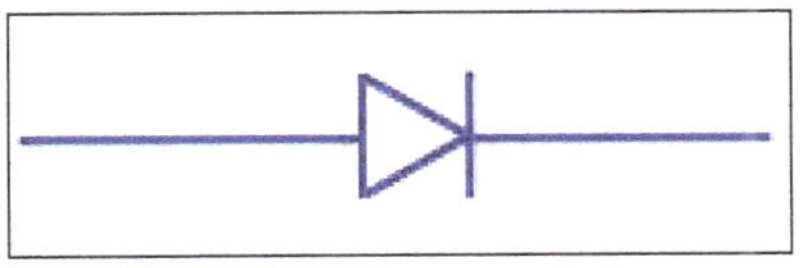

Diode circuit symbol

2. Light Emitting Diode (LED): LED is used to emit light when a current is passed through the device. It is abbreviated as LED.

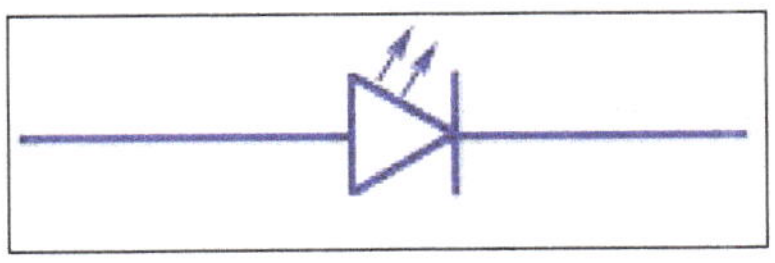

LED Circuit symbol

3. Zener Diode: After a breakdown voltage, the device allows current to flow in the reverse direction as well.

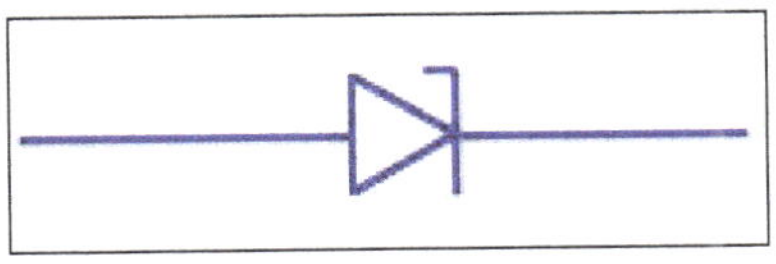

Zener diode circuit symbol

4. Photo Diode: Photodiode works as a photo-detector and converts light into its corresponding voltage or current.

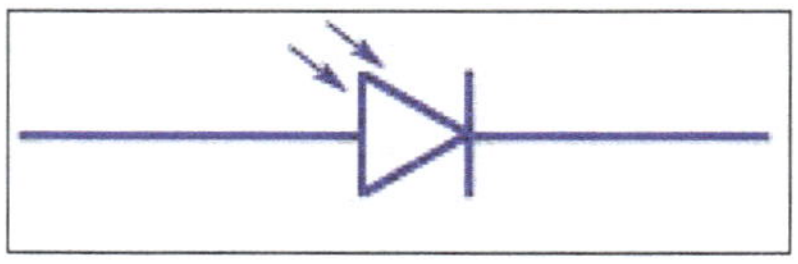

Photo diode circuit symbol

5. Tunnel Diode: Tunnel Diode is known for its high-speed operation due to its application in quantum mechanical effects.

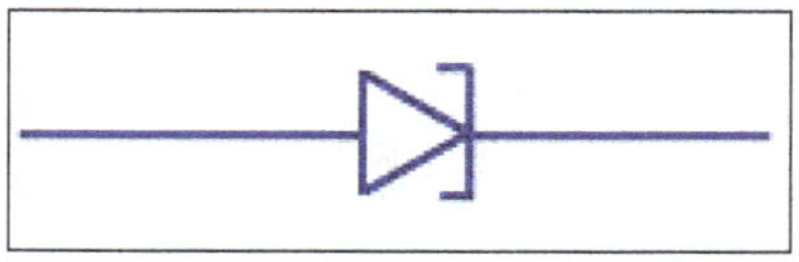

Tunnel diode circuit symbol

6. Schottky Diode: The Schottky Diode is known for its large forward voltage drop and hence has great applications in switching circuits.

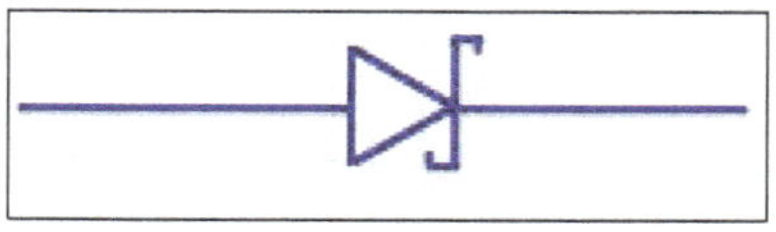

Schottky diode circuit symbol

Transistor Symbols

1. NPN Transistor: This is a transistor with a layer of P-doped semiconductor fixed between two layers of N-doped semiconductors that act as the emitter and collector.

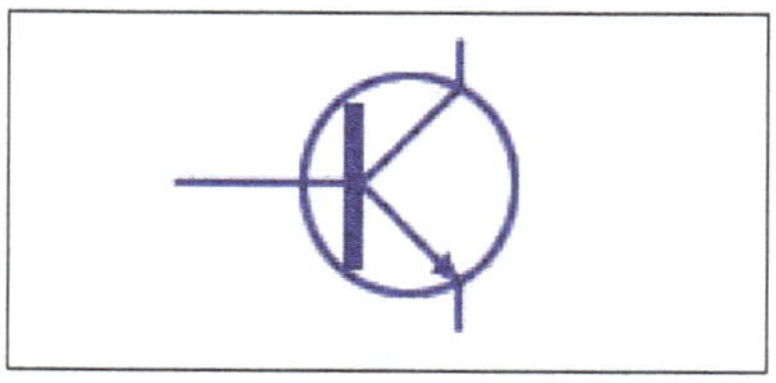

Transistor NPN circuit symbol

2. PNP Transistor: This is a transistor with a layer of N-doped semiconductor fixed between two layers of P-doped semiconductors that act as the emitter and collector.

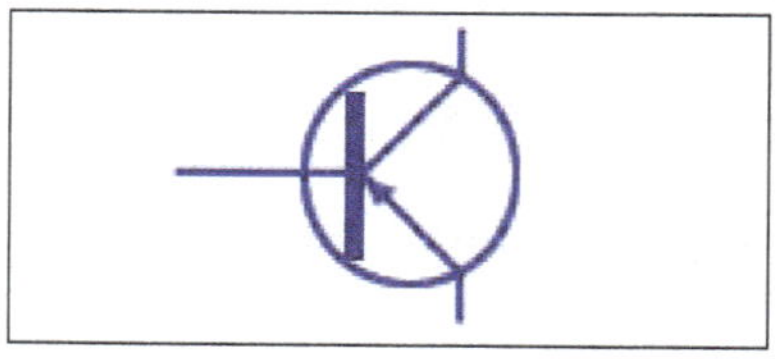

Transistor PNP circuit symbol

3. Phototransistor: The working of a phototransistors similar to that of a bipolar transistor with a difference that it converts light into its corresponding current. The phototransistor can also act as a photodiode if the emitter is not connected.

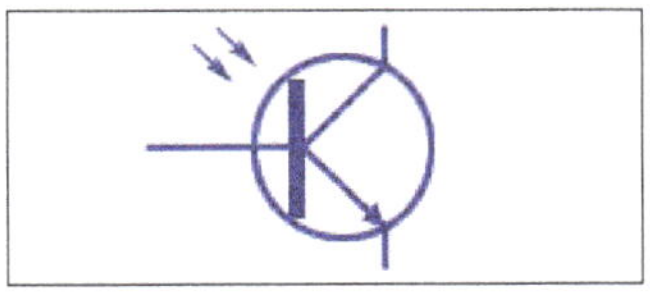

Phototransistor circuit symbol

4. Field Effect Transistor: Like a transistor, a FET has three terminals, the Gate, Source and Drain. The device has an electric field that controls the conductivity of a channel of one type charge carrier in a semiconductor substance.

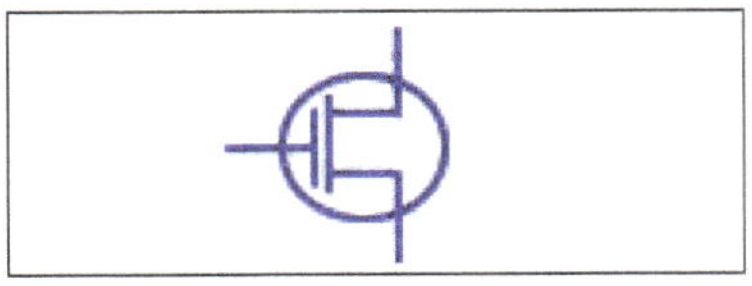

Field effect transistor circuit symbol

5. N-Channel Junction FET: The Junction Field Effect Transistor (JFET) is the simplest type of FET with applications in Switching and voltage variable resistor. In an N-channel JFET an N-type silicon bar has two smaller pieces of P-type silicon material diffused on each sides of its middle part, forming P-N junctions.

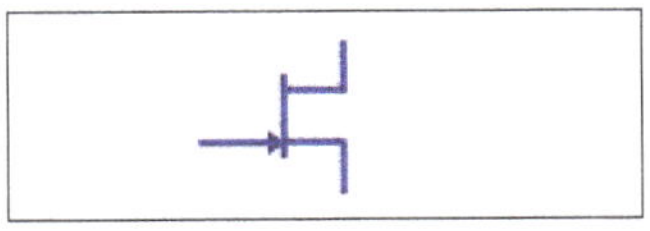

n-channel junction field effect transistor (JFET) circuit symbol

6. P-Channel Junction FET: P-channel JFET is similar in construction to N-channel JFET except that P-type semiconductor base is sandwiched between two N-type junctions. In this case majority carriers are holes.

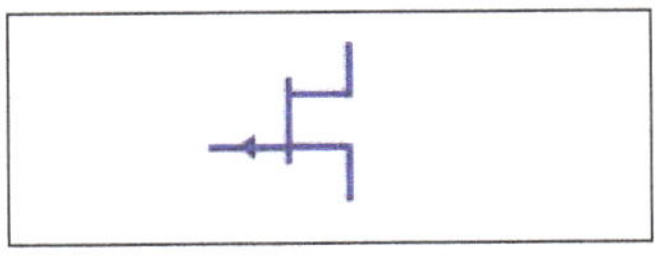

p-channel junction field effect transistor (FET) circuit symbol

7. Metal Oxide Semiconductor FET: Abbreviated as MOSFET. MOSFET is a three terminal device and is controlled by a gate bias. It is known for its low capacitance and low input impedance.

8. Enhancement MOSFET: The enhancement MOSFET structure has no channel formed during its construction. Voltage is applied to the gate, so as to develop a channel of charge carriers so that a current results when a voltage is applied across the drain-source terminals.

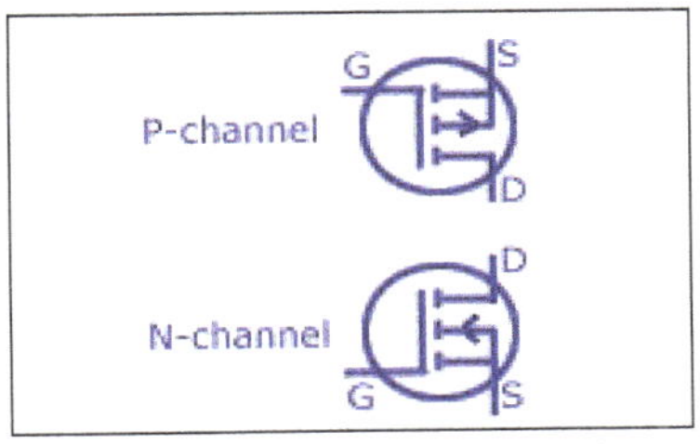

e-MOSFET circuit symbol

9. Depletion MOSFET: In the depletion-mode construction a channel is physically constructed and a current between drain and source is due to voltage applied across the drain-source terminals.

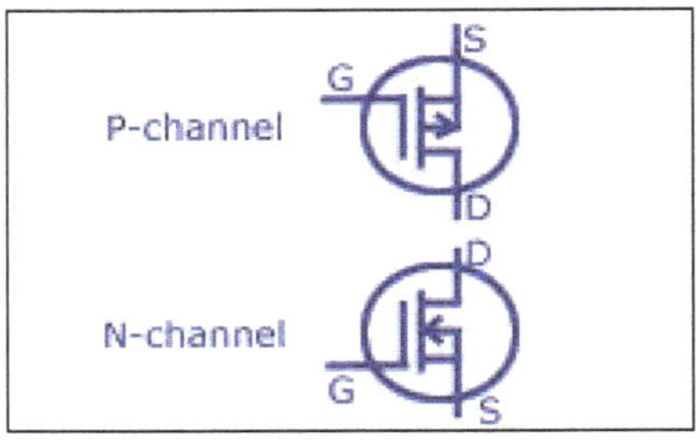

d-MOSFET circuit symbol

Logic Gate Symbols

1. AND Gate: If all the inputs of an AND gate are HIGH, then the output will also be HIGH. If any one of them is LOW, the output will also be LOW.

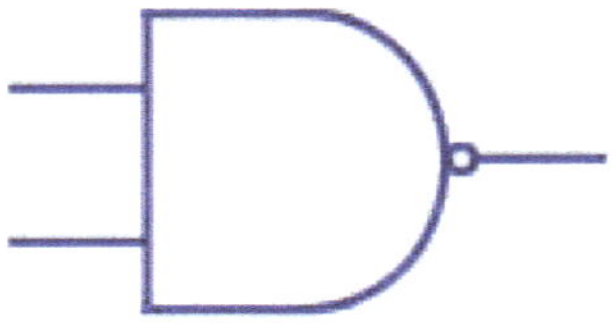	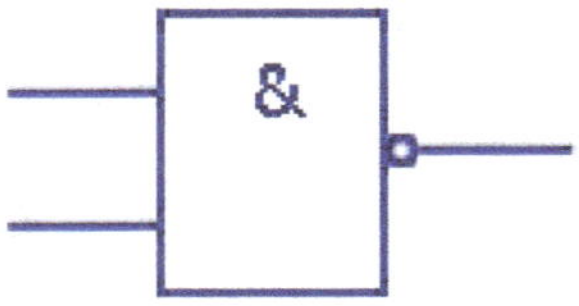
AND Gate IEC Symbol	AND GATE Symbol

2. NAND Gate : Short form for NOT AND Gate. Of all the inputs are HIGH, the output will be LOW. If any one input is LOW, the output will be HIGH.

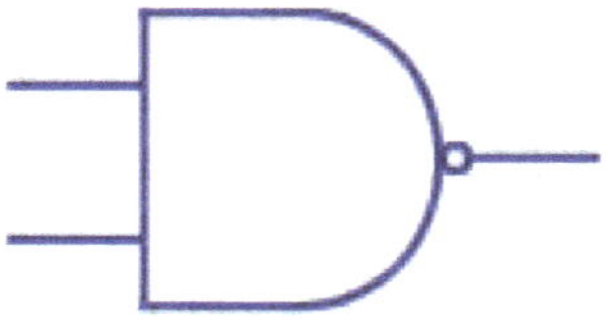	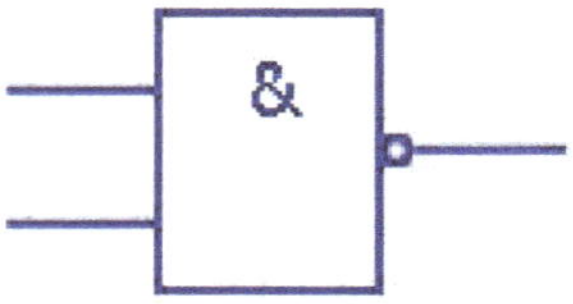
NAND Gate Symbol	NAND Gate IEC Symbol

3. OR Gate: If any one of the input is HIGH, the output will also be HIGH. If both inputs are LOW, the output will also be LOW.

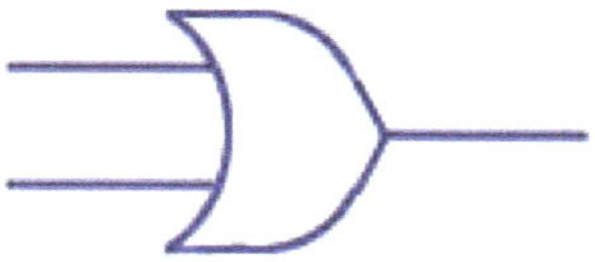	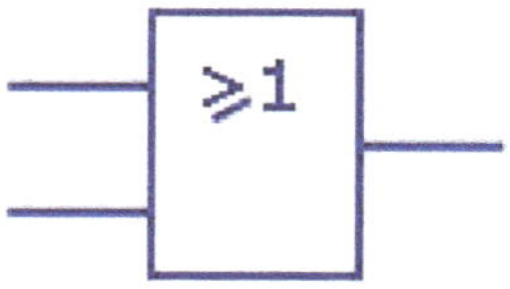
OR Gate Symbol	OR Gate IEC Symbol

4. NOR Gate: Short form for NOT OR. If both inputs are LOW, the output will also be LOW. For other cases, the output will be HIGH.

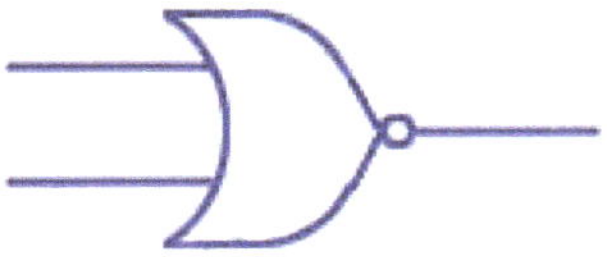	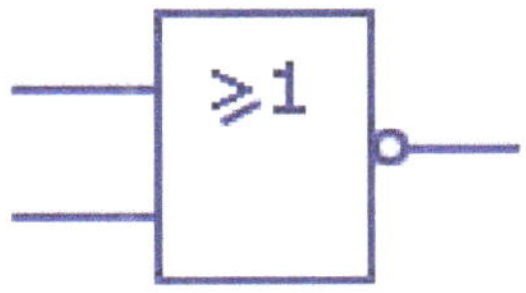
NOR Gate Symbol	NOR Gate IEC Symbol

5. EX-OR Gate: Short form for Exclusive NOT OR. If both the inputs are the same, the output will be HIGH. If both are different, the output will also be different.

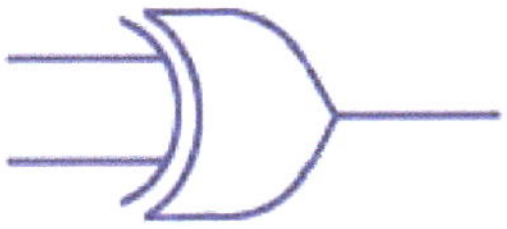	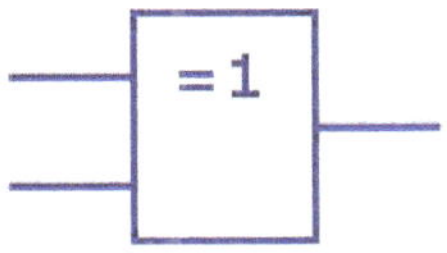
EX-OR Gate Symbol	EX-OR Gate IEC Symbol

6. EX-NOR Gate: Short form for Exclusive NOT OR. If both the inputs are the same, the output will be HIGH. If both are different, the output will also be different.

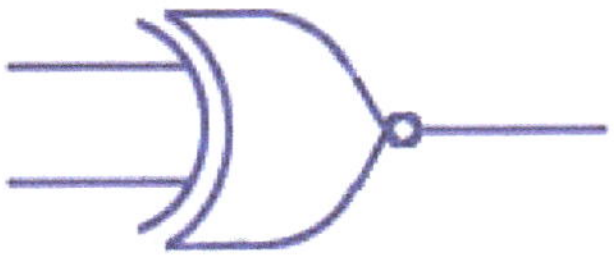	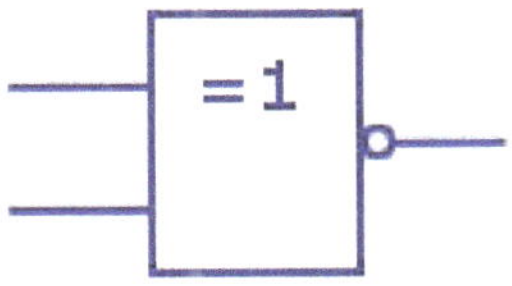
EX-NOR Gate Symbol	EX-NOR Gate IEC Symbol

7. NOT Gate: Also known as the inverter Gate. There is only one input for this gate. If the input is HIGH, the output will be LOW. If the input is LOW, the output will be HIGH.

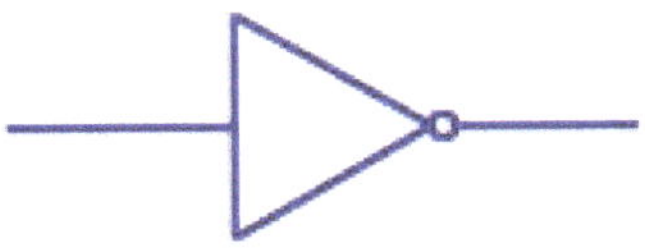	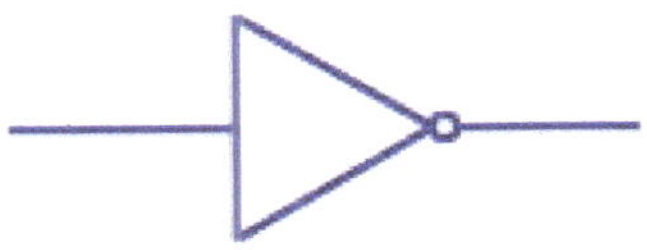
NOT Gate Symbol	NOT Gate Symbol

Meters

1. Voltmeter: Voltmeter is used to measure the voltage at a certain point in the circuit.

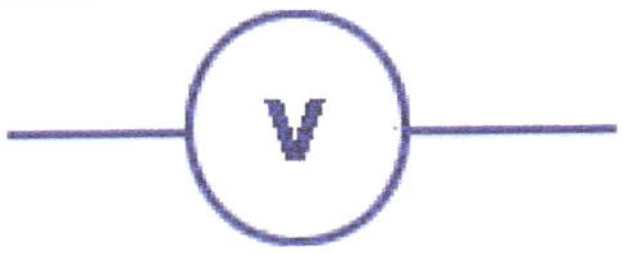

Voltmeter Circuit Symbol

2. Ammeter: An Ammeter is used to measure the current that passes through the circuit at a particular point.

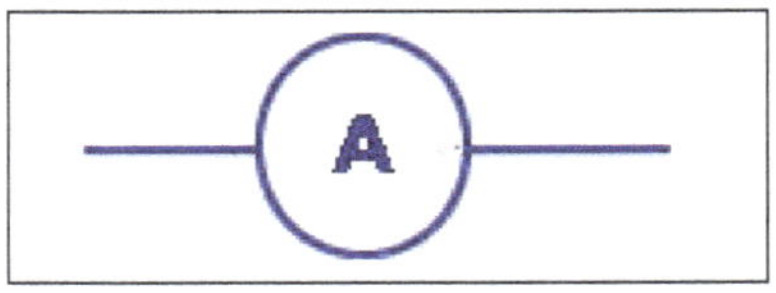

Ammeter circuit symbol

3. Galvanometer: A galvanometer is used to measure very small currents in the order of 1 milli ampere or less.

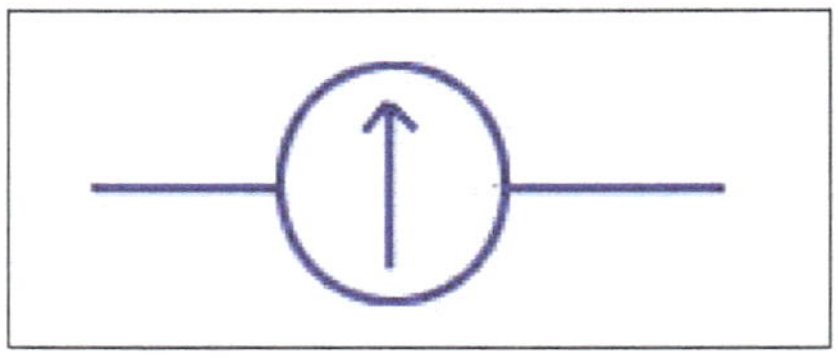

Galvanometer circuit symbol

4. Ohmmeter: Resistance of the circuit is measured using an Ohmmeter.

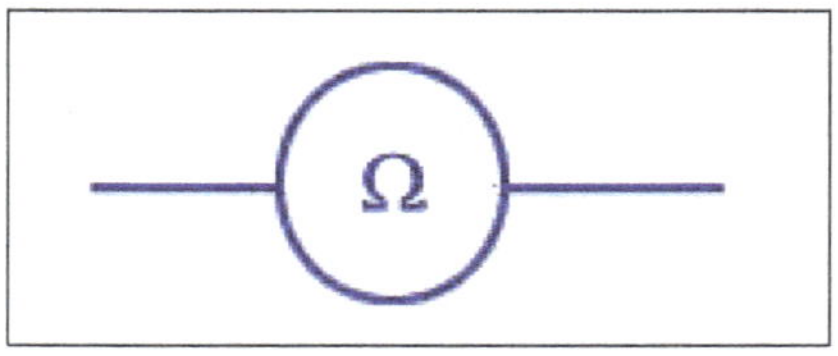

Ohmmeter circuit symbol

5. Oscilloscope: An oscilloscope is used to measure the voltage and time period of signals along with their shape display.

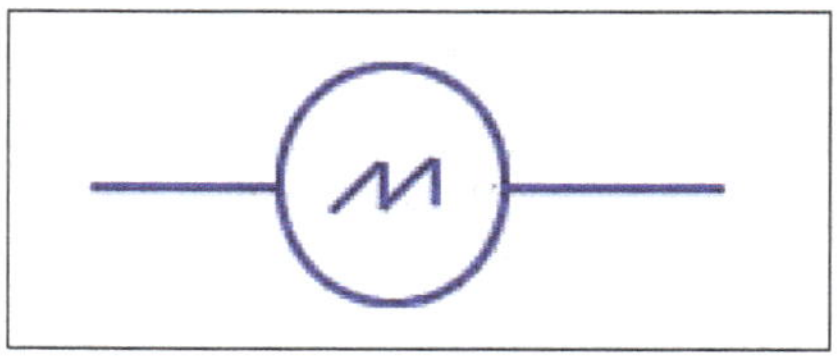

Oscilloscope circuit symbol

Sensor Symbols

1. Light Dependent Resistor (LDR): It is abbreviated as LDR. Light Dependent Resistor is used to convert light into its corresponding resistance. Instead of directly measuring the light, it senses the heat content and converts it onto resistance.

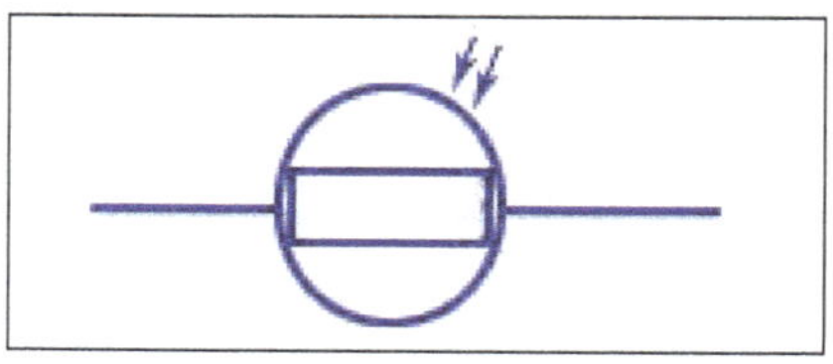

LDR circuit symbol

2. Thermistor: Instead of directly measuring the light, a thermistor senses the heat content and converts it into resistance.

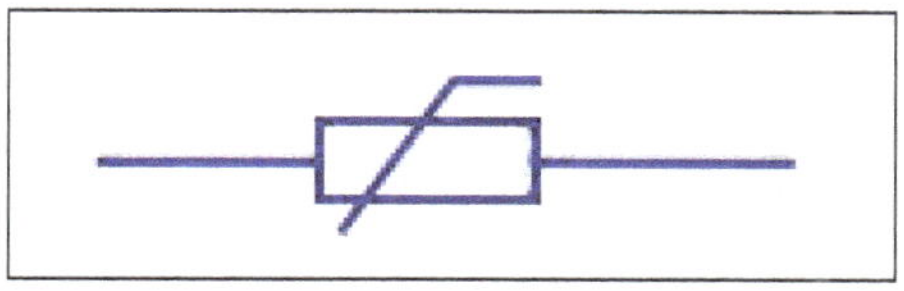

Thermistor circuit symbol

Switch Symbols

1. Push Switch: This is an ordinary switch that passes current only upon pressing.

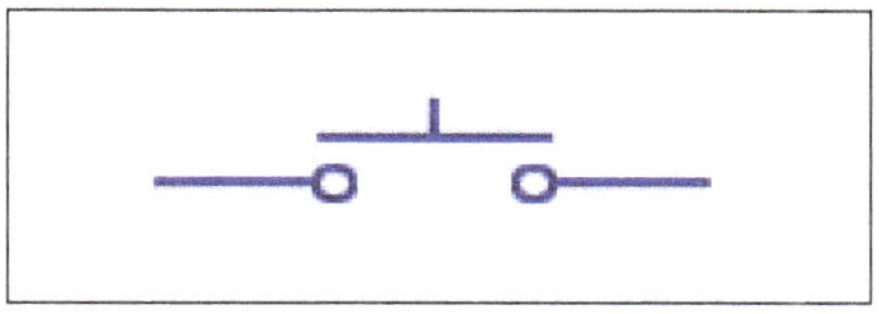

Push switch circuit symbol

2. Push to Break Switch: The push to break switch is usually kept in the ON state (closed). It turns to OFF state (open) only when the switch is pressed.

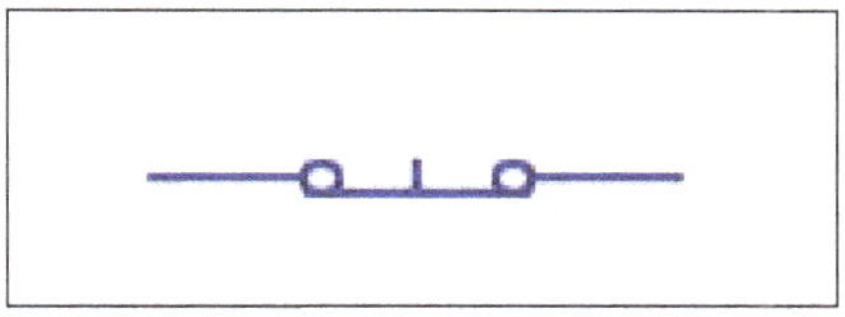

Push to break switch circuit symbol

3. Singe Pole Single Throw Switch: Also known as the ON/OFF switch. This switch allows the flow of current only when it is kept ON.

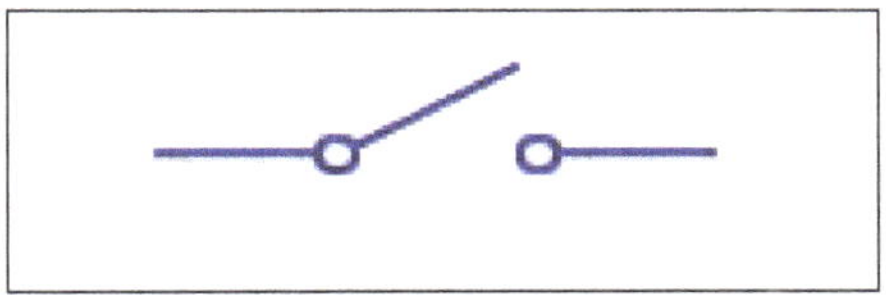

On off switch (SPST) circuit symbol

4. Single Pole Double Throw Switch: Also known as the 2-way switch. It can be also called as an ON/OFF/ON switch as it has an OFF position in the center. The switch causes the flow of current in two directions, depending on its position. It can be abbreviated as SPDT.

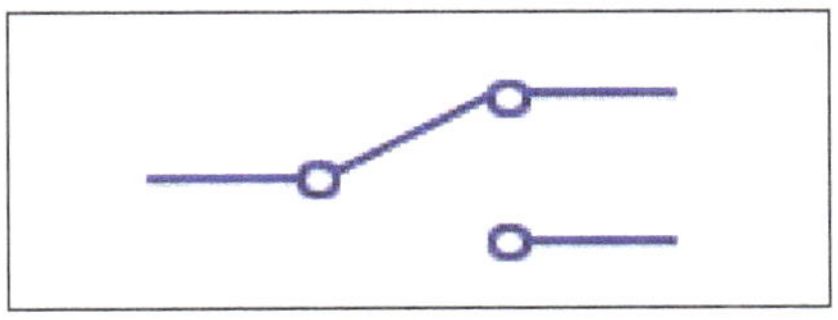

2-way switch (SPDT) circuit symbol

5. Double Pole Single Throw Switch: Abbreviated as DPST. Can also be called as a dual ON-OFF switch. This is used to isolate between the live and neutral connections in the main electrical line.

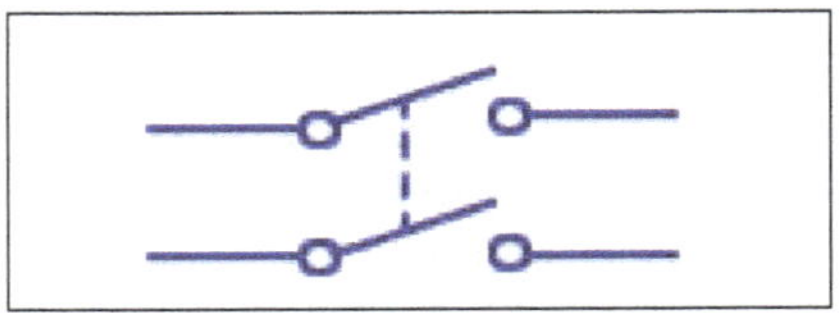

Dual on-off switch (DPST) circuit symbol

6. Double Pole Double Throw Switch: Abbreviated as DPDT. The switch uses a central OFF position and is applied as reversing switch for motors.

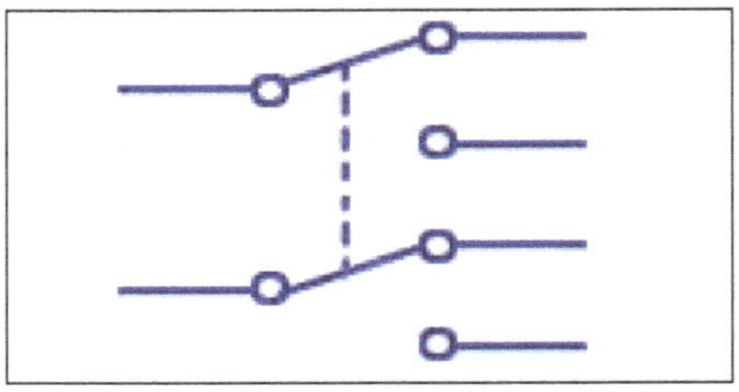

DPDT circuit symbol

7. Relay: This device can easily switch a 230 Volt AC mains circuit. It has three switching stages called Normally Open (NO). Normally Closed (NC), and Common (COM).

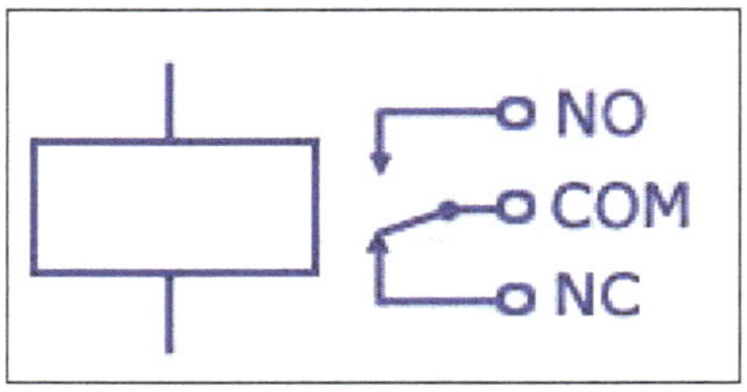

Relay circuit symbol

Audio and Radio Device Symbols

1. Microphone: This device is used for converting sound to its corresponding electrical energy.

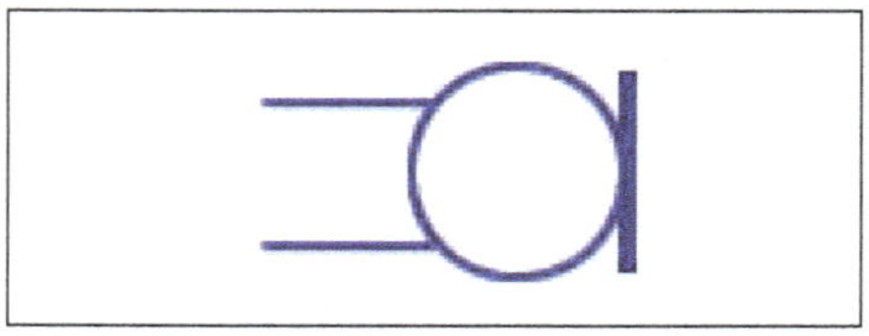

Microphone circuit symbol

2. Earphone: Does the reverse process of microphone and converts electrical energy into sound.

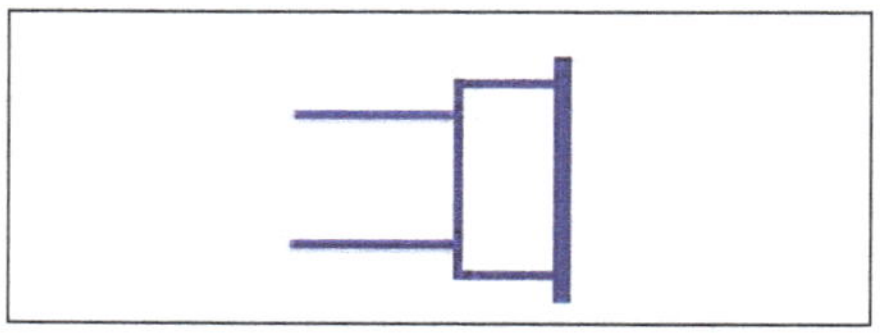

Earphone circuit symbol

3. Loudspeaker: Does the same operation as an earphone, but converts an amplified version of the electrical energy into its corresponding sound.

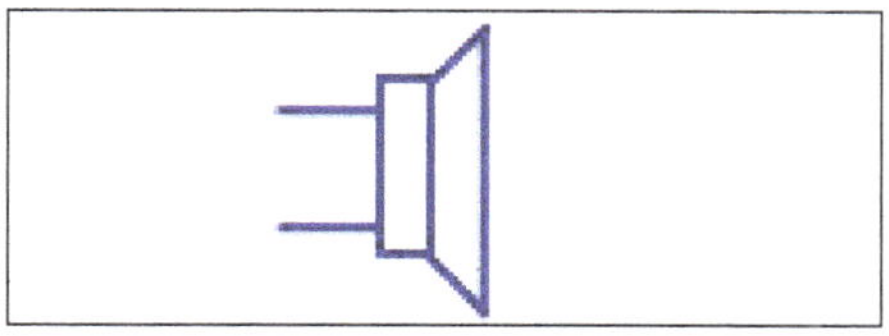

Loudspeaker circuit symbol

4. Piezo-Transducer: It is a transducer that converts electrical energy into sound.

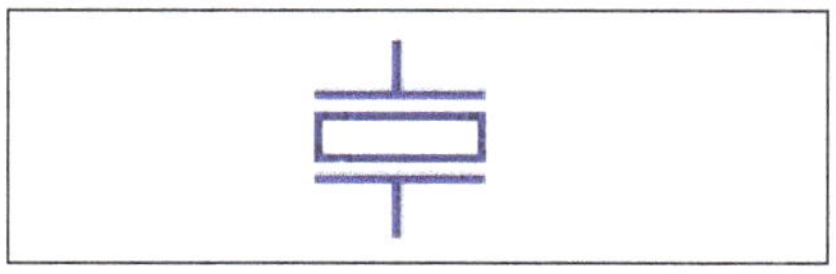

Piezotransducer circuit symbol

5. Amplifier": Used to amplify a signal. It is mainly used to represent a whole circuit rather than just one component.

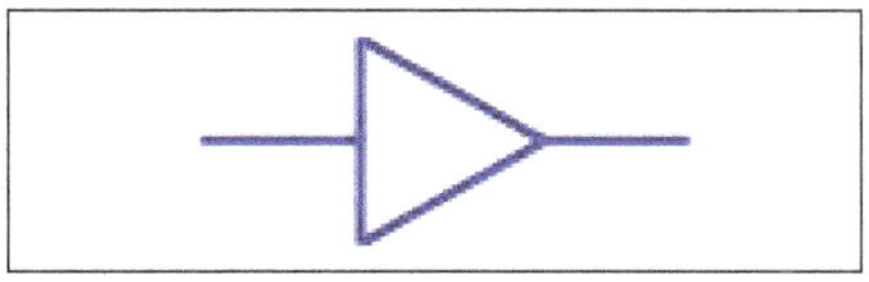

Amplifier circuit symbol

6. Aerial: This device is used to transmit/receive signals.

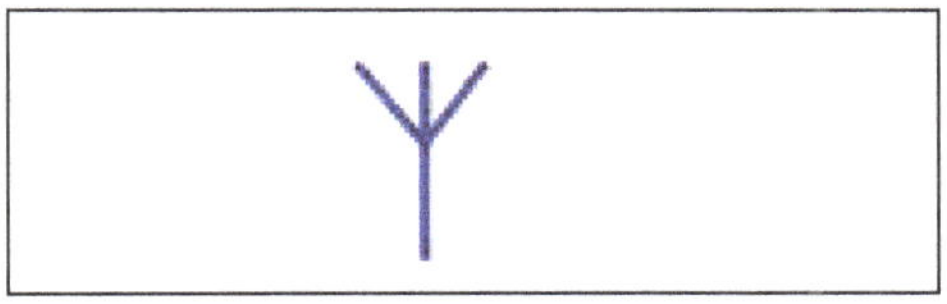

Aerial circuit symbol

Output Devices

1. Lighting Lamp: This is used to provide light for the output.

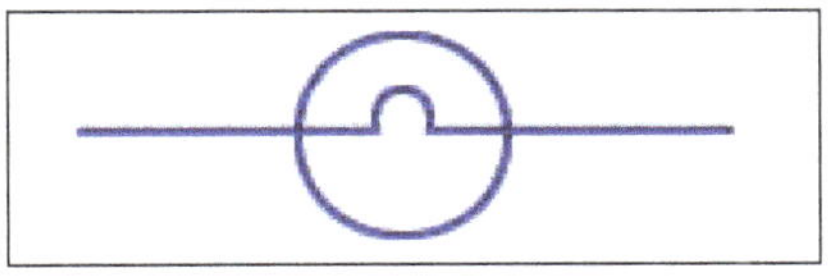

Lamp circuit symbol

2. Indicator Lamp: Used to convert electrical energy into light. The best example is the warning light on a car dashboard.

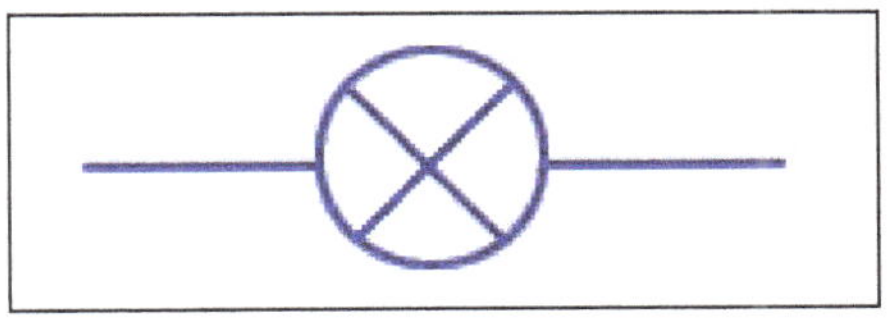

Lamp indiator circuit symbol

3. Heater: This transducer is used to change electrical energy into heat.

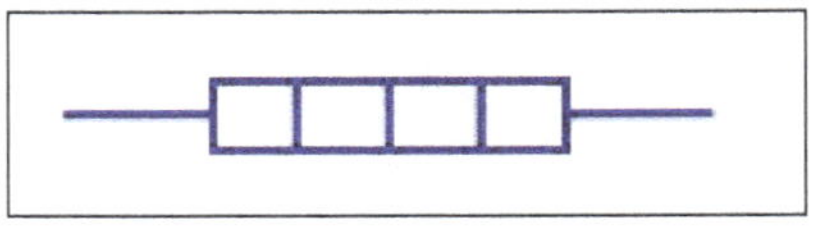

Heater circuit symbol

4. Inductor: Inductor is used to produce a magnetic field when a certain current is passed through a coil of wire. The wire is coiled on a soft iron core. Have applications in motors, and tank circuits.

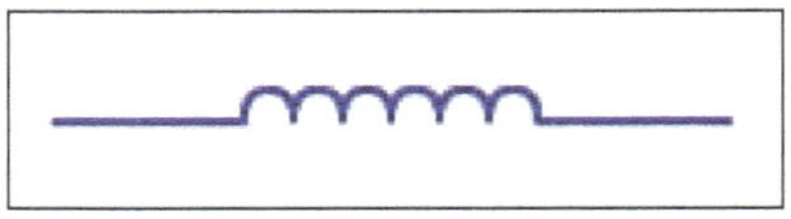

Inductor circuit symbol

5. Motor: This device is used to convert electrical energy into mechanical energy. Can be used as a generator as well.

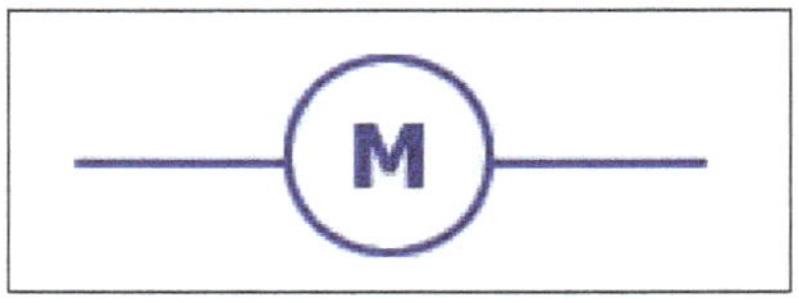

Motor circuit symbol

6. Bell: Used to produce a sound as the output, according to the electrical energy produced as the input.

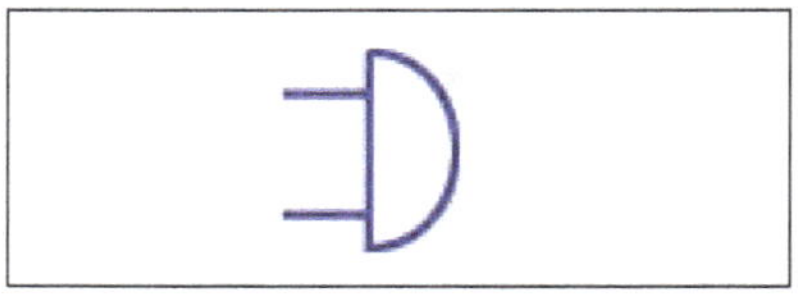

Bell circuit symbol

7. Buzzer: It is used to produce an output sound corresponding to the electrical energy in the input.

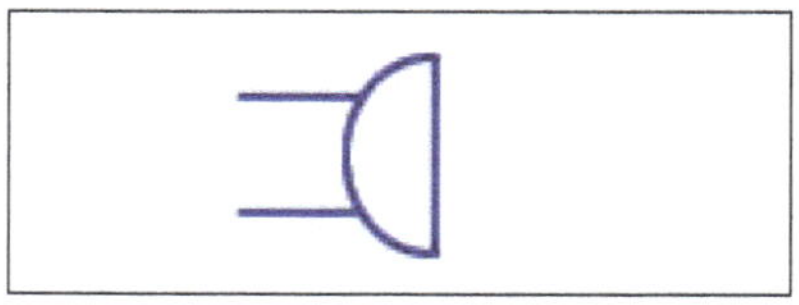

Buzzer circuit symbol

Printed Circuit Board

PCB is a copper laminated and non-conductive Printed Circuit Board, in which all electrical and electronic components are connected together in one common board with physical support for all components with base of board. When PCB is not developed, at that time all components are connected with a wire which increases complexity and decreases reliability of the circuit, by this way we cannot make a very large circuit like motherboard. In PCB, all components are connected without wires, all components are connected internally, so it will reduce the complexity of the overall circuit design. PCB is used to provide electricity and connectivity between the components, by which it functions the way it was designed. PCBs can be customised for any specifications to user requirements. It can be found in many electronics devices like; TV, Mobile, Digital camera, Computers parts like; Graphic cards, Motherboard, etc. It also used in many fields like; medical devices, industrial machinery, automotive industries, lighting, etc.

Types of PCB

There are several types of PCB available for the circuit. Out of these types of PCB, we have to choose the appropriate type of PCB according to our application.

- Single-layer PCB
- Double-layer PCB
- Multi-layer PCB
- Flexible PCB
- Aluminium backed PCB
- Flex-rigid PCB

Single Layer PCB

A single layer PCB is also known as single sided PCB. This type of PCB is simple and most used PCB because these PCBs are easy to design and manufacture. One side of this PCB is coated with a layer of any conducting material. Generally, copper is used as conducting material for PCB, because copper has very good conducting characteristic. A layer of solder mask is used to protect PCB against oxidation followed by silk screen to mark out all of the components on the PCB. In this type of PCB, only one side of the PCB is used to connect different types of electrical or electronics components like resistor, capacitor, inductor, etc. These components are soldered. These PCBs are used in low cost and bulk manufacturing application like calculators, radio, printers and the solid-state drive.

Single layer PCB

Double Layer PCB

Double layer PCB is also known as double sided PCB. As name suggests, in this type of PCB, a thin layer of conducting material, like copper is applied to both top and bottom sides of the board. In PCB, on different layer of board, consist via, which has two pads in corresponding position on different layers. These are electrically connected by a hole through the board, which is shown in figure-2b. More flexible, relatively lower cost, and most important advantage of this type of PCB board is its reduced size which makes circuit compact. This type of PCB is mostly used in industrial controls, converter, UPS system, HVAC application, Phone, Amplifier and Power monitoring systems.

Double layer PCB

Multilayer PCB

Multilayer PCB has more than two layers. It means that, this type of PCB has at least three conductive layers of copper. For securing the board glue is sandwiched between the layer of insulation which ensures that the excess heat will not damage any component of circuit. This type PCB designing is very complex and used in very complicated and large electrical task in very low space and compact circuit. This type of PCB is used in large application like GPS technology, satellite system, medical equipment, file server and data storage.

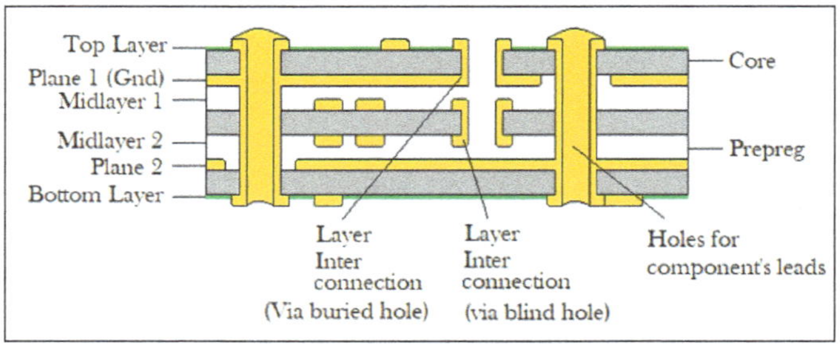

Multi-layer PCB

Flexible PCB

Flexible PCB is also known as Flex circuit. This type of PCB used flexible plastic material like polymide, PEEK (Polyether ether ketone) or transparent conductive polyester film. The circuit board is generally place in folded or twisted. This is very complex type of PCB and it also contains different range of layers like single sided flex circuit, double sided flex circuit and multisided flex circuit. Flex circuit is used in organic light emitting diode, LCD fabrication, flex solar cell, automotive industries, cellular telephones, camera and complex electronics devices like laptops.

Flexible PCB

Rigid PCB

Rigid PCBs are made out of solid material which don't allow PCB from twisting. Same as flex PCB

Rigid PCB also have different layer configuration like single layer, double layer and multi-layer Rigid PCB. Shape of this PCB is not change after installation. This PCB cannot be bent according to shape of base that's why this PCB is known as RIGID PCB. Lifespan of this type of PCB is very high, so this is used in many parts of computer like RAM, GPU and CPU. Simple in design and most used and most manufacture PCB is single sided rigid PCB. Multi-layer rigid PCB can more compact by containing 9-10 layers.

Flex-rigid PCB

Combination of Flexible circuit and rigid circuit is most important board. Flex-rigid boards consists of multiple layers of flexible PCB attached to a number of rigid PCB layer. Flex-rigid board is as shown in figure. It is used in cell phones, digital cameras and automobiles etc.

Flex rigid PCB

Types of PCBs According to Mounting System:

1. Through-hole PCB.
2. Surface mounted PCB.

Through-hole PCB

In this type of PCB, we have to make hole using drill on PCB. In these holes, leads of components are mounted and soldered to pads situated on opposite side of PCB. This technology is most useful because it gives more mechanical support to electrical components and very reliable technology for mounting of components but drilling in PCB make it more expensive. In single layer PCB, this mounting technology is easy to implement, but in case of double layer and multi-layer PCB making hole is more difficult.

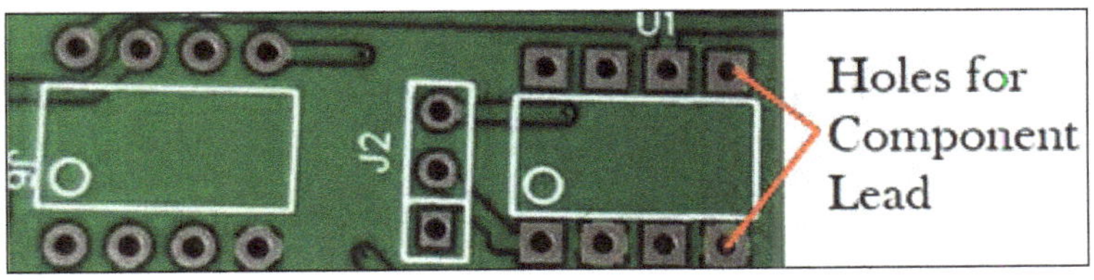

Through-hole PCB

Surface Mounted PCB

In this type of PCB, components are small in size because these components have very small lead or no leads are required for mounting on the board. Here, in this technology, SMD components are directly mounted on the surface of the board and not require to make hole on board.

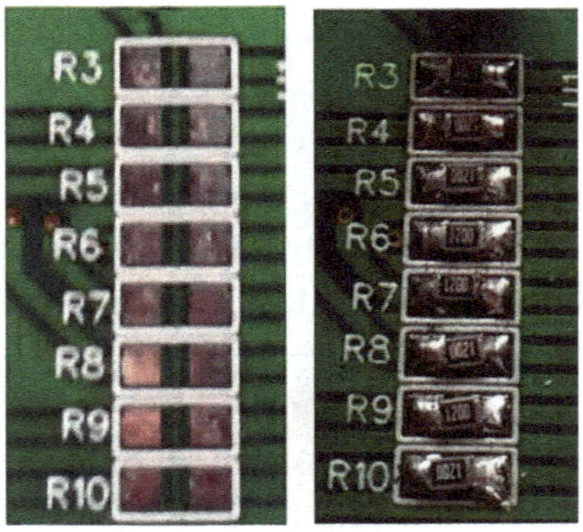

Surface mounted PCB

Different Parts of PCB

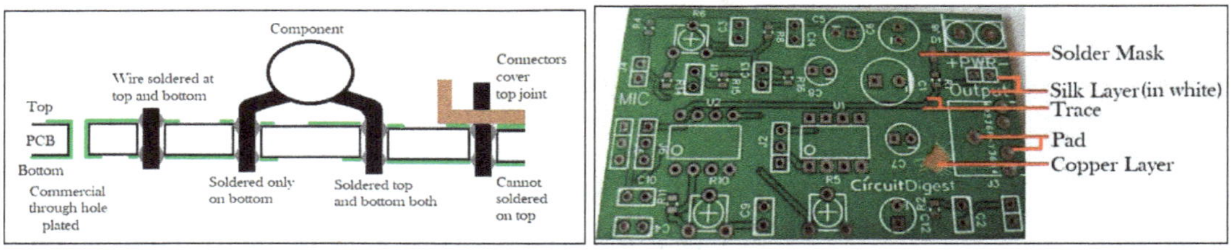

- Pad: Pad is nothing but a piece of copper on which lead of components are mounted and on which soldering are done. Pad provides the mechanical support to the components.

- Trace: In PCB, components are not connected with the help of wires. All components are connected with a conducting material like copper. This copper part of PCB which is used to connect all components that is known as trace. Trace is looks like below figure.

- Layers: According to application, cost and available space of circuit, user can choose the layer of PCB. Most simple in construction, easy to design and most useful in routine life is single layer PCB. But for very large and complex circuit, double layer PCB or Multi-layer PCB is most preferred compared to single layer PCB. Now a day, in multi-layer PCB, 10-12 layers can be connected and most critical thing is to communicate between the components in different layer.

- Silk layer: Silk layer is used for printing line, text or any art on the surface of PCB. Usually, for screen printing epoxy ink is used. Silk layer can be used in top and/or bottom layer of PCB according to user requirement which is known as silk screen TOP and silk screen BOTTOM.

- Top and bottom layer: In Top layer of PCB, all components are mounted in this layer of PCB. Generally, this layer is green coloured. In bottom layer of PCB, all components are soldered through the hole and lead of components is known as bottom layer of PCB. Sometime, in top and bottom layer PCB is coated with green colour layer, which is known as solder mask.

- Solder Mask: There is one additional layer on the top of copper layer called as Solder Mask. This layer generally has green color but it can be of any color. This insulating layer is used for to prevent accidental contact of pads with other conductive material on PCB.

PCB Materials

The main element is dielectric substrate which is rigid or flexible. This dielectric substrate is used with conducting material like copper on it. As dielectric material, the glass epoxy laminates or composite materials are used.

FR4

FR is stand for FIRE RETARDENT. For all type of PCB manufacturing, most common glass laminated material is FR4. Based on woven glass-epoxy compounds, FR4 is a composite material which is most useful because it provides very good mechanical strength.

FR4	Glass transition temp.	Most common and widely used
Standard	130	Cheapest
With higher glass transition temp.	170-180	Compatible with the lead free reflow technology
Halogen free	--	Compatible with the lead free reflow technology

FR-1 and FR-2

This material is made from paper and phenol compounds and this material is used for only single layer PCB. Both FR1 and FR2 has similar characteristic, the only difference is glass transition temperature. FR1 has higher glass transition temperature compared to FR2. These materials is also sub divided in standard, halogen free and non-hydrophobic.

CEM-1

These material is made from paper and two layer of woven glass epoxy and phenol compounds and this material is used for Single sided PCB only. CEM-1 can used instead of FR4, but price of CEM1 is higher than FR4.

CEM-3

This material is white coloured, glass epoxy compound which is mostly used in double layer PCB. CEM-3 has lower mechanical strength compared to FR4, but it is cheaper than FR4. So, this is a good alternative of FR4.

Polyimide

This material is used in flexible PCB. This material is made from kepton, rogers, dupont. This material has good electrical properties, felicity, wide temperature range and high chemical resistance. Working temperature of this material is -200 °C to 300 °C.

Prepreg

Prepreg means pre-impregnated. It is a Fiberglass impregnated with resin. These resins are pre-dried, so that when it heated, it flows, sticks and completely immersed. Prepreg has adhesive layer which gives strength similar to FR4. There are many versions of this material according to resin content, SR- standard resin, MR- medium resin and HR- high resin. This is chosen according to required thickness, layer structure and impedance. This material also available in high glass transition temperature and halogen free.

Electrical Circuit

Electrical circuit is an interconnection of electrical components. An electrical circuit consists of batteries, resistors, inductors, capacitors, switches or transistors. An electrical network consists of a closed loop. A circuit is a closed path where electrons flow in a wire. As long as the copper wire is allowed to itself, the electrons drift between the atoms but never leave the copper.

However, when we connect this copper wire to a battery the free electrons will be driven towards the positive terminal of the battery. This pushing force is called Electromotive force (E.M.F). The E.M.F. is expressed in volts. And usually, it is called voltage. As a result of this voltage, an electron motion takes place. This motion is known as electron current or electric current. We can measure current by connecting an ammeter between copper wire and voltage source.

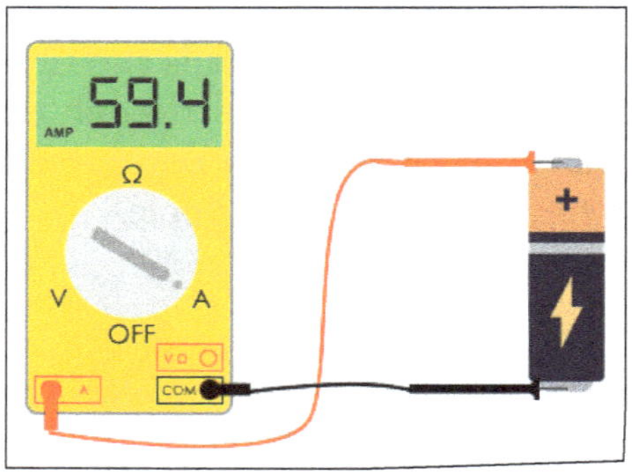

A complete circuit is a never-ending loop of electrons. If we take a wire and loop it around, it forms a continuous path in which electrons can flow forever. This is a primary concept of a circuit.

An electric circuit mainly consists of:

- Electrical sources that provide voltage and current like batteries. They are the source of electrons.
- Switches, resistors, potentiometers, capacitors which are used to control electricity.
- Protection devices in high voltage circuits. They are MCB, fuse, etc.

- Wires which are conducting path to carry electric current from one point to another in a circuit.
- Load in the circuit can be motor, LED, lamp, etc.

There are some basic properties of electrical circuits and they are:

- The circuit is always a closed path.
- A circuit always consists of an energy source.
- Direction of flow of current is from positive terminal to negative terminal of the source.
- Direction of flow of electrons is from negative terminal to positive terminal of the source.

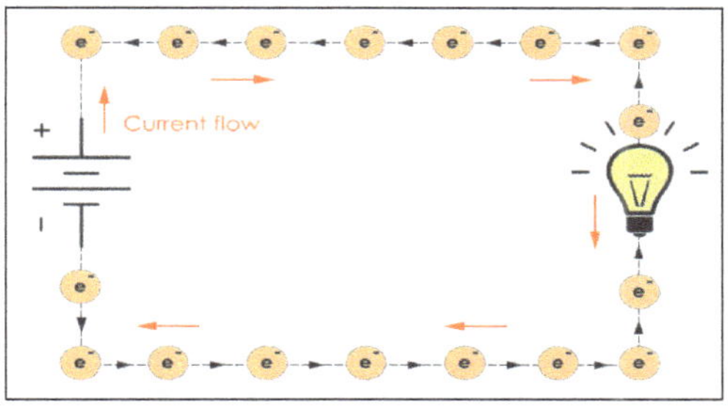

Circuit Diagram

A circuit diagram is a visual display of an electrical circuit. There are mainly two types of circuit diagrams:

- Pictorial: Pictorial diagrams are made using basic images. This type of diagram gives a visual representation to audience which is less technical.

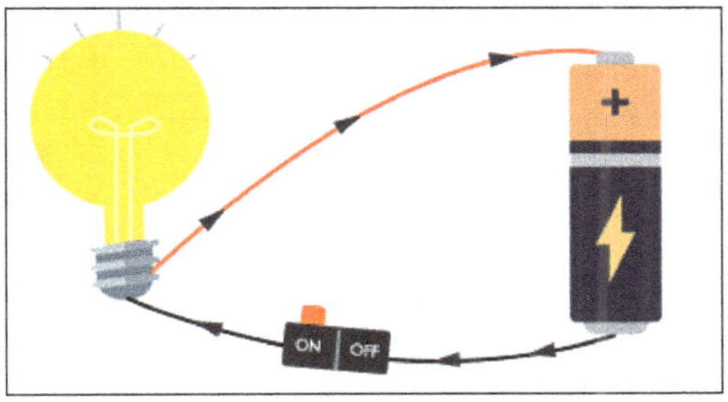

A pictorial circuit diagram

- Schematic: These diagrams are drawn using standard industrial symbols. These diagrams are used for the representation of a circuit to an electrician or any other technical audience.

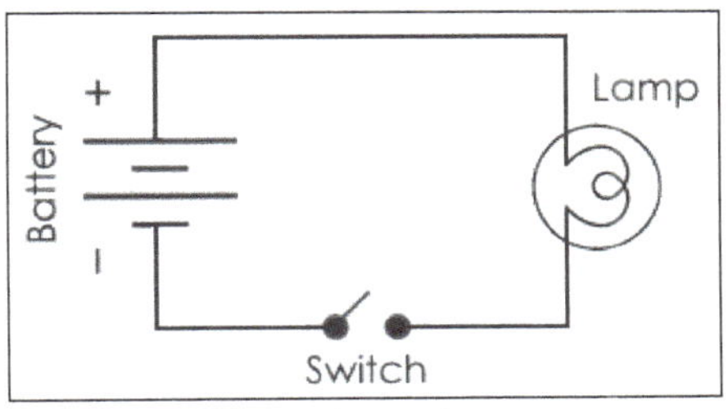

A schematic

Circuit Diagram Symbols

There are hundreds of symbols for the circuit diagram. Some basic symbols are:

Name	**Symbol**
Resistor	
Capacitor	
Cell	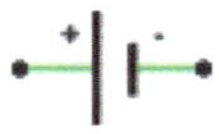
Battery	
LED	

Suppose we want to draw a simple circuit in which battery is connected to LED in such a way that positive terminal of the battery is connected to positive terminal of LED and negative terminal of the battery is connected to negative terminal of LED. Then it can be represented as:

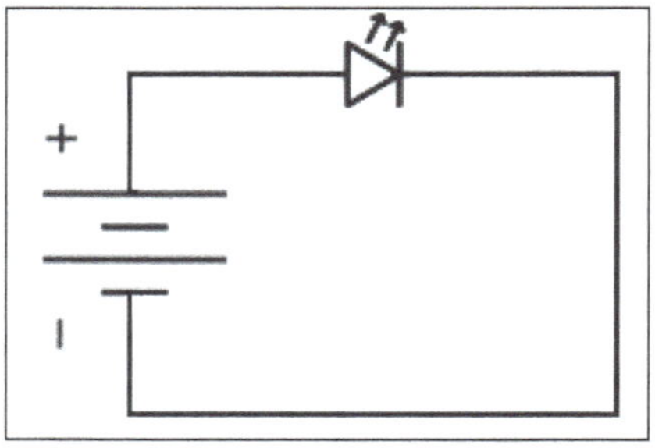

Types of Circuits

There are basically three types of circuits:

Open Circuit

If in a simple circuit one terminal is disconnected, then there is no flow of current through that circuit. This is said to be an open circuit or no load condition.

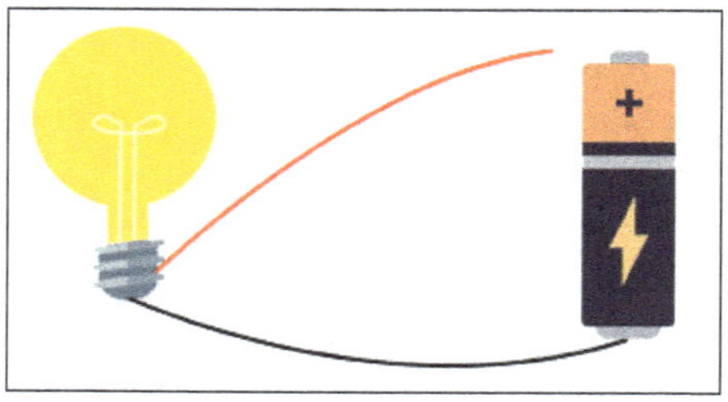

An open circuit

Closed Circuit

An electric circuit has a source of Electromotive force and a load. This load acts as a conductor path. If the current flows through the load it is considered as a closed circuit. If in a simple circuit, current can flow from one terminal of the battery to another without any discontinuation is said to be closed circuit.

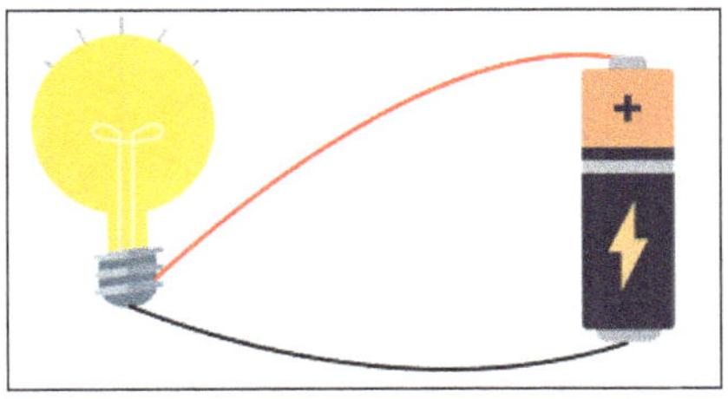

A closed circuit

Short Circuit

If the positive terminal of the battery is directly connected to the negative terminal without any resistance between them is said to be a short circuit.

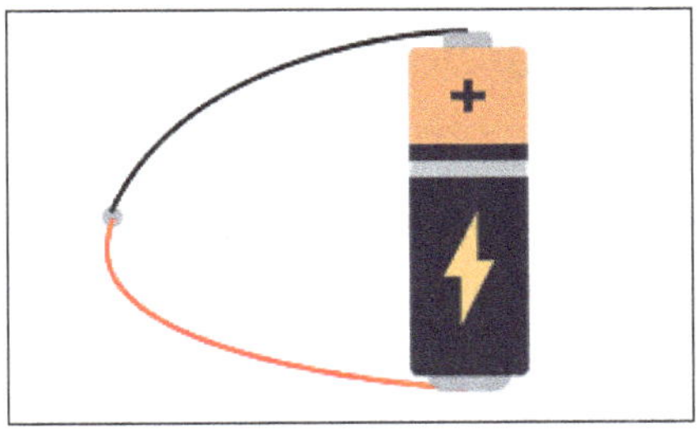

A Short Circuit

Other than the above circuits, components in an electric circuit can be arranged in two different ways and that are in series connection and in parallel connection.

Series Circuit

If in a circuit, components are connected in series then the circuit is known as a series circuit. In a series circuit, current through each component is same and voltage supplied is the sum of the voltage across each component. If a wire joins the battery to one lamp, to the next lamp and then back to the battery, the lamps are said to be connected in series.

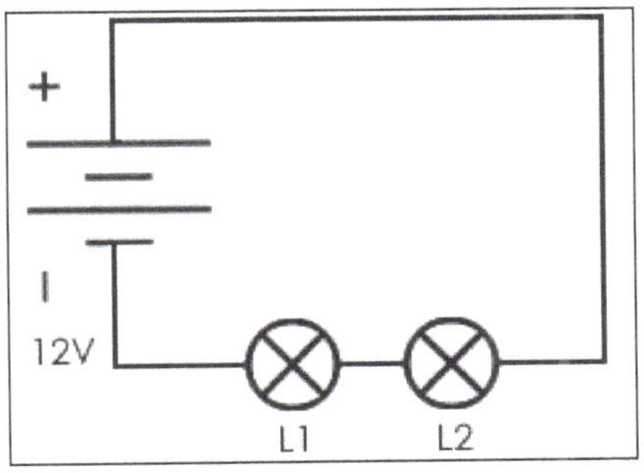

Series connection of two lamps

Parallel Circuit

If in a circuit, components are connected in parallel then the circuit is known as parallel circuit. In a parallel circuit, the voltage across each component will be same and total current applied is the sum of current through each component. If a lamp is connected to the battery and another lamp is connected in a separate loop with first lamp, then lamp is connected in parallel connection.

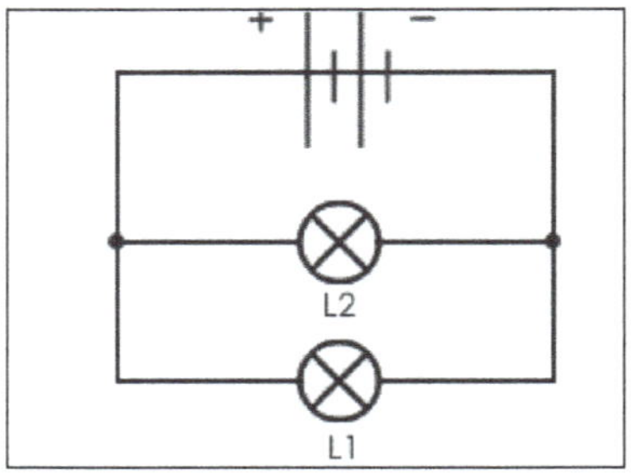

Parallel connection of two lamps

Here, the voltage across each bulb would be the same as the voltage applied by the battery. Current across each lamp would be divided means if we apply 5A to the circuit, 5A will be the current flowing through each lamp.

Thus this is how series and parallel circuits work and they have their own current and voltage dividing properties.

Electric circuits are all around us, in our mobile phones, in our computers, in fan and also in the torch. It is difficult to assume the practical use of electricity without circuits. We are all dependent on these complex circuits around us.

DC Circuit

The closed path in which the direct current flows is called the DC circuit. The current flows in only one direction and it is mostly used in low voltage applications. The resistor is the main component of the DC circuit.

A simple DC circuit is shown in the figure below which contains a DC source (battery), a load lamp, a switch, connecting leads, and measuring instruments like ammeter and voltmeter. The load resistor is connected in series, parallel or series-parallel combination as per requirement.

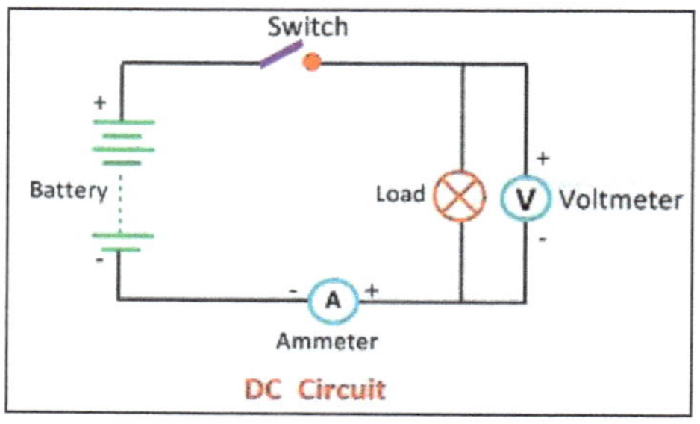

Types of DC Circuit

The DC electric circuit is mainly classified into three groups. They are the series DC circuit, parallel DC circuit, and series and parallel DC circuit.

DC Series Circuit

The circuit in which have DC series source, and the number of resistors are connected end to end so that same current flow through them is called a DC series circuit. The figure below shows the simple series circuit. In the series circuit the resistor R_1, R_2, and R_3 are connected in series across a supply voltage of V volts. The same current I is flowing through all the three resistors.

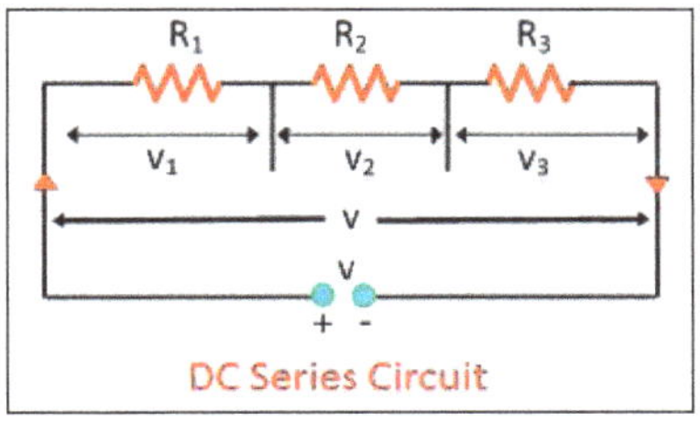

If V_1, V_2, and V_3 are the voltage drop across the three resistor R_1, R_2, and R_3 respectively, then

$$V = V_1 + V_2 + V_3$$
$$V = IR_1 + IR_2 + IR_3$$

Let R be the total resistance of the circuit then,

$$IR = IR_1 + IR_2 + IR_3$$
$$R = IR_1 + IR_2 + IR_3$$

Total resistance = Sum of the individual resistance

In such type of circuit all the lamps are controlled by the single switch and they cannot be controlled individually. The most common application of this circuit is for decoration purpose where a number of low voltage lamps are connected in series.

DC Parallel Circuit

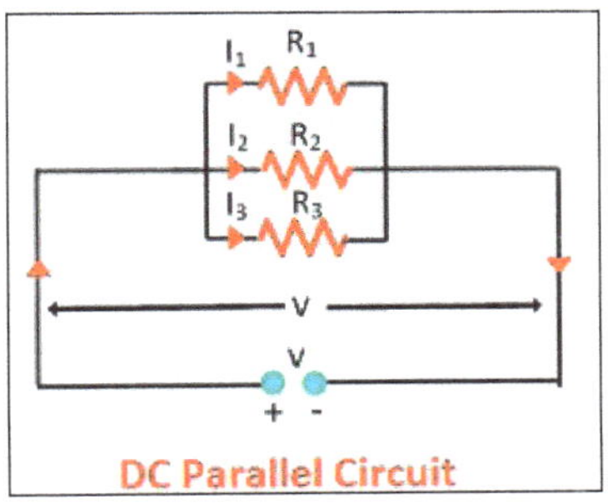

The circuit which have DC source and one end of all the resistors is joined to a common point and other end are also joined to another common point so that current flows through them is called a DC parallel circuit.

The figure shows a simple parallel circuit. In this circuit the three resistor R_1, R_2, and R_3 are connected in parallel across a supply voltage of V volts. The current flowing through them is I_1, I_2 and I_3 respectively.

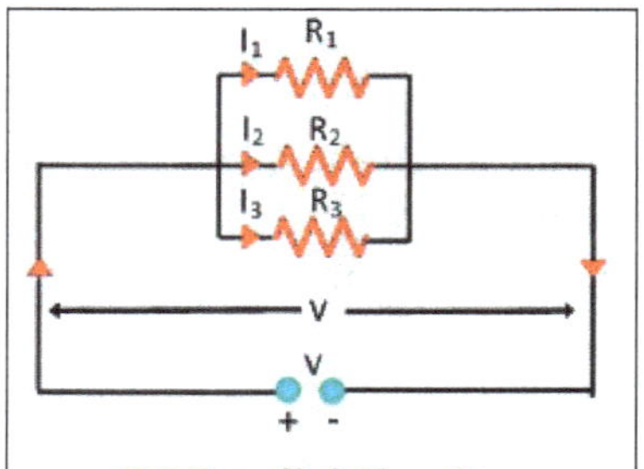

DC Parallel Circuit

The total current drawn by the circuit

$$I = I_1 + I_2 + I_3$$

$$I = \frac{V}{R_1} + \frac{V}{R_2} + \frac{V}{R_3}$$

Let R be the total or effective resistance of the circuit, then

$$\frac{V}{R} = \frac{V}{R_1} + \frac{V}{R_2} + \frac{V}{R_3}$$

$$\frac{1}{R} = \frac{1}{R_1} + \frac{1}{R_2} + \frac{1}{R_3}$$

Reciprocal of total resistance = sum of reciprocal of the individual resistance.

All the resistance is operated to the same voltage therefore all of them are connected in parallel. Each of them can be controlled individually with the help of a separate switch.

DC Series-parallel Circuit

The circuit in which series and parallel circuit are connected in series is called a series parallel circuit. The figure below show the series-parallel circuit. In this circuit, two resistor R_1 and R_2 are connected in parallel with each other across terminal AB. The other three resistors R_3, R_4 and R_6 are connected in parallel with each other across terminal BC.

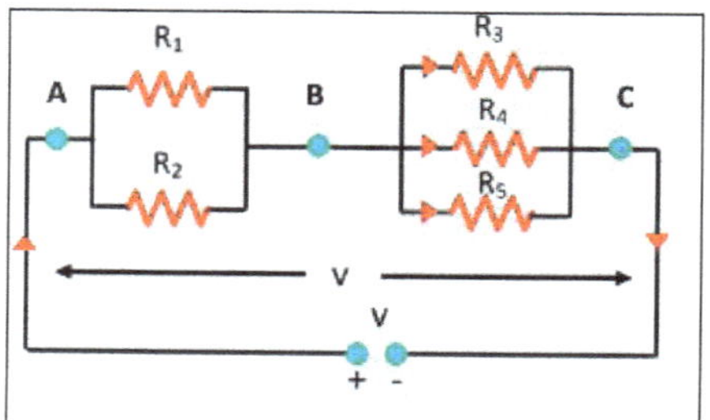

DC Series Parallel Circuit

The two groups of resistor R_{AB} and R_{BC} are connected in series with each other across the supply voltage of V volts. The total or effective resistance of the whole circuit can be determined as given below,

$$\frac{1}{R_{AB}} = \frac{1}{R_1} + \frac{1}{R_2} + \frac{R_1 + R_2}{R_1 R_2}$$

$$R_{AB} = \frac{R_1 R_2}{R_1 + R_2}$$

Similarly,

$$\frac{1}{R_{BC}} = \frac{1}{R_3} + \frac{1}{R_4} + \frac{1}{R_5} = \frac{R_3 + R_4 + R_4 + R_5 + R_5 + R_3}{R_3 R_4 R_5}$$

$$R_{BC} = \frac{R_3 R_4 R_5}{R_3 + R_4 + R_4 + R_5 + R_5 + R_3}$$

Total or effective resistance of the ciruit,

$$R = R_{AB} + R_{BC}$$

AC Circuit

AC circuits as the name (Alternating Current) implies are simply circuits powered by an Alternating Source, either voltage or current. An Alternating Current or Voltage, is one in which the value of either the voltage or the current varies about a particular mean value and reverses direction periodically.

Most present day household and industrial Appliances and systems are powered using alternating current. All DC based plugged in appliances and rechargeable battery based devices technically run on Alternating current as they all use some form of DC power derived from AC for either charging of their batteries or powering of the system. Thus Alternating current is the form via which power is delivered at the mains.

Resistance in an AC circuit

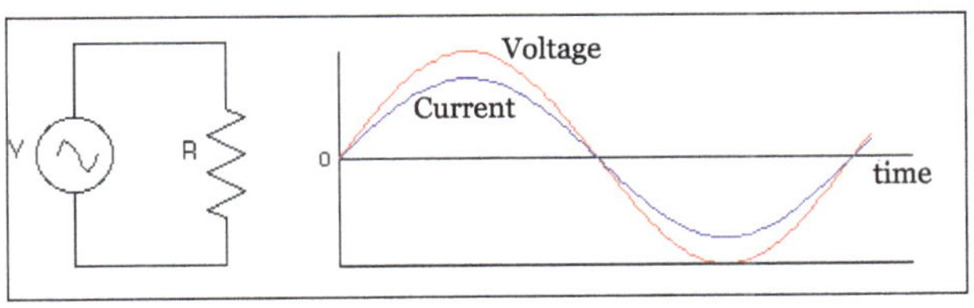

The relationship V = IR applies for resistors in an AC circuit, so

$$I = V/R = (V_0/R)\sin(\omega t) = I_0 \sin(\omega t)$$

In AC circuits we'll talk a lot about the phase of the current relative to the voltage. In a circuit which only involves resistors, the current and voltage are in phase with each other, which means that the peak voltage is reached at the same instant as peak current. In circuits which have capacitors and inductors (coils) the phase relationships will be quite different.

Capacitance in an AC Circuit

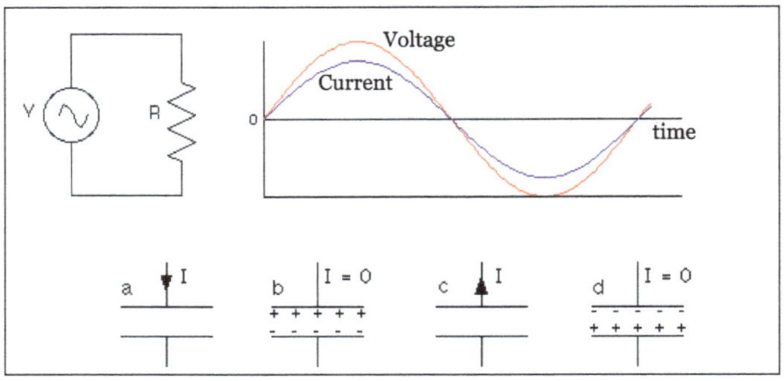

Consider now a circuit which has only a capacitor and an AC power source (such as a wall outlet). A capacitor is a device for storing charging. It turns out that there is a 90° phase difference between the current and voltage, with the current reaching its peak 90° (1/4 cycle) before the voltage reaches its peak. the current leads the voltage by 90° in a purely capacitive circuit.

To understand why this is, we should review some of the relevant equations, including:

Relationship between voltage and charge for a capacitor: CV = Q

Relationship between current and the flow of charge,

$$I = \Delta Q / \Delta t$$

The AC power supply produces an oscillating voltage. We should follow the circuit through one cycle of the voltage to figure out what happens to the current.

Step 1 - At point a the voltage is zero and the capacitor is uncharged. Initially, the voltage increases quickly. The voltage across the capacitor matches the power supply voltage, so the current is large to build up charge on the capacitor plates. The closer the voltage gets to its peak, the slower it changes, meaning less current has to flow. When the voltage reaches a peak at point b, the capacitor is fully charged and the current is momentarily zero.

Step 2 - After reaching a peak, the voltage starts dropping. The capacitor must discharge now, so the current reverses direction. When the voltage passes through zero at point c, it's changing quite rapidly; to match this voltage the current must be large and negative.

Step 3 - Between points c and d, the voltage is negative. Charge builds up again on the capacitor plates, but the polarity is opposite to what it was in step one. Again the current is negative, and as the voltage reaches its negative peak at point d the current drops to zero.

Step 4 - After point d, the voltage heads toward zero and the capacitor must discharge. When the voltage reaches zero it's gone through a full cycle so it's back to point a again to repeat the cycle.

The larger the capacitance of the capacitor, the more charge has to flow to build up a particular voltage on the plates, and the higher the current will be. The higher the frequency of the voltage, the shorter the time available to change the voltage, so the larger the current has to be. The current, then, increases as the capacitance increases and as the frequency increases.

Usually this is thought of in terms of the effective resistance of the capacitor, which is known as the capacitive reactance, measured in ohms. There is an inverse relationship between current and resistance, so the capacitive reactance is inversely proportional to the capacitance and the frequency:

A capacitor in an AC circuit exhibits a kind of resistance called capacitive reactance, measured in ohms. This depends on the frequency of the AC voltage, and is given by:

$$\text{Capacitive reactance: } X_c = 1/\omega C = 1/2\pi f C$$

We can use this like a resistance (because, really, it is a resistance) in an equation of the form V = IR to get the voltage across the capacitor:

$$V = IX_c$$

V and I are generally the rms values of the voltage and current.

Inductance in an AC circuit

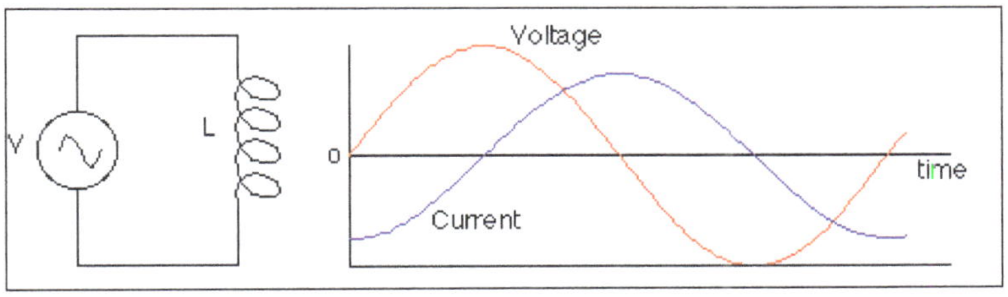

An inductor is simply a coil of wire (often wrapped around a piece of ferromagnet). If we now look at a circuit composed only of an inductor and an AC power source, we will again find that there is a 90° phase difference between the voltage and the current in the inductor. This time, however, the current lags the voltage by 90°, so it reaches its peak 1/4 cycle after the voltage peaks.

The reason for this has to do with the law of induction:

$$\varepsilon = {-N\Delta\phi}/{\Delta t} \text{ or } \varepsilon = {-L\Delta T}/{\Delta t}$$

Applying Kirchoff's loop rule to the circuit above gives:

$$V - {L\Delta T}/{\Delta t} = 0 \text{ or } V = {L\Delta I}/{\Delta t}$$

As the voltage from the power source increases from zero, the voltage on the inductor matches it. With the capacitor, the voltage came from the charge stored on the capacitor plates (or, equivalently, from the electric field between the plates). With the inductor, the voltage comes from changing the flux through the coil, or, equivalently, changing the current through the coil, which changes the magnetic field in the coil.

To produce a large positive voltage, a large increase in current is required. When the voltage passes through zero, the current should stop changing just for an instant. When the voltage is large and negative, the current should be decreasing quickly. These conditions can all be satisfied by having the current vary like a negative cosine wave, when the voltage follows a sine wave.

If the frequency is raised, there is less time to change the voltage. If the time interval is reduced, the change in current is also reduced, so the current is lower. The current is also reduced if the inductance is increased.

As with the capacitor, this is usually put in terms of the effective resistance of the inductor. This effective resistance is known as the inductive reactance. This is given by:

$$X_L = \omega L = 2\pi fL$$

Where L is the inductance of the coil (this depends on the geometry of the coil and whether its got a ferromagnetic core). The unit of inductance is the henry.

As with capacitive reactance, the voltage across the inductor is given by:

$$V = IX_L.$$

RC Circuit

The combination of a pure resistance R in ohms and pure capacitance C in Farads is called RC circuit. The capacitor store energy and the resistor connect in series with the capacitor control the charging and discharging of a capacitor. The RC circuit is used in camera flashes, pacemaker, timing circuit etc.

The RC signal filters the signals by blocking some frequencies and allowing others to pass through it. It is also called first order RC circuit and is used to filter the signals by passing some frequencies and blocking others. The RC fitlers are mostly use for selecting signals and for rejecting noise.

The high-pass filter and the low-pass filter are the most common type of RC filters. The high pass filter passes the frequency greater than the fixed cutoff frequency and blocks the frequency lower than the fixed cutoff frequency. Similarly, the low-pass filter allows the frequency lower than the fixed cutoff frequency and attenuates the frequency higher than the fixed cutoff frequency.

Series RC Circuits

When components R and C are in series, the current must be the same (magnitude and phase) in all parts of the circuit. The current through the resistor R is in phase with the voltage across R. But the capacitive current is out of phase with the capacitor voltage: I_c leads V_c by 90°. Therefore, the voltages across R and C are out of phase.

Each component, R and C, has its own series voltage drop (I_R and I_x). These voltages are also out of phase with each other, up to 90°. As shown in figure below, the sum of the individual voltage drops appears to be more than the applied voltage V_T, but this is due to their phase difference. However, the vector sum is equal to V_T.

The series current can be calculated using Ohm's law: $I_T = V_{drop} / X_c$ and R. However, when a sine wave is applied to an RC series circuit, the total opposition to current flow is really a combination of R and X_C. This total opposition is known as impedance, symbolized by Z. To solve for the total circuit impedance, use the same formula that is used to find the hypotenuse of a right triangle.

$$Z = \sqrt{R^2 + X_C{}^2}$$

Knowing Z and the total applied voltage, you can solve for the total circuit current I_T using Ohm's law where $I_T = V_T / Z$

Also, by knowing the values of X_c and R, the circuit impedance angle (with respect to circuit current *I)* can be calculated as the inverse tangent of the ratio X_c / R (the negative sign means the voltage lags current):

$$\text{Series RC impendance angle}, \theta_z = \tan^{-1} - Xc^{IR}$$

This will indicate if the circuit is more capacitive ($X_c > R$) or resistive ($R > X_c$) A phase angle of -45 degrees means that R and X_c are equal. However, if R is close to 0 and X_c is high, then the phase angle resembles a purely capacitive circuit and is close to -90 degrees. In general, a practical rule is that when X_c or R is 10 times greater than the other, the circuit takes on that characteristic. So, if R is 1 kilohm and X_c is 100 ohms, then the circuit is resistive and the phase angle is closer to 0. In other words, there is little or no phase shift between voltage and current figure below shows the impedance of a series RC circuit where X_C and R is equal values.

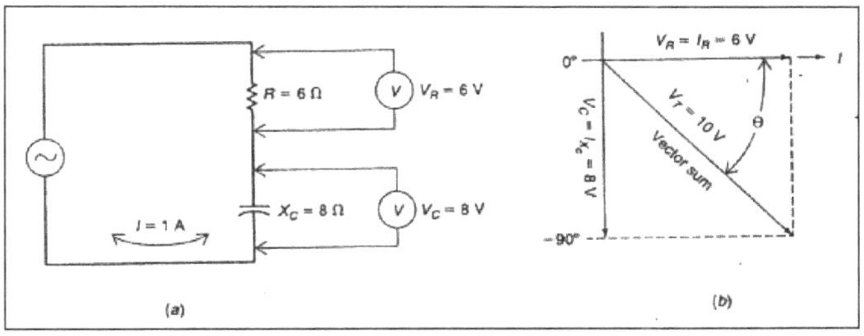

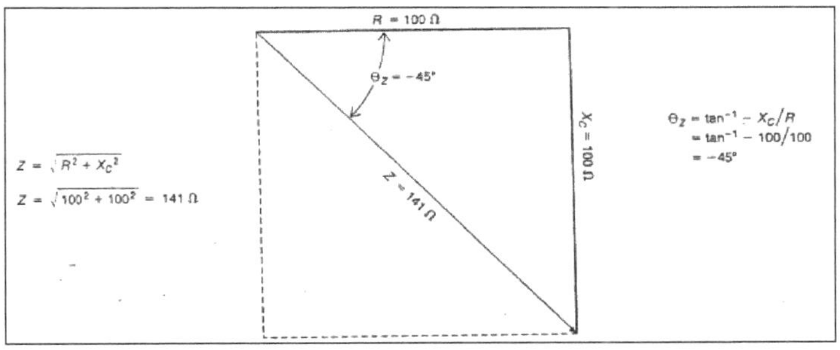

Parallel RC Circuits

When R and C are in parallel, each branch has its own value of current depending upon the branch value of R or X_c. Also, because they are in parallel, the voltage (magnitude and phase) across the parallel branches is the same. Therefore, there is no phase difference between the applied voltage and the voltage across R and C in parallel.

The current through the resistive branch is in phase with the applied signal. But the current through the capacitive branch leads its voltage V_c by 90 degrees. Therefore, the two branch currents (I_r and Ic) are out of phase (90 degrees apart). With equal values of R of R and X_c, the total current in the main line is 45" apart from either branch' figure below shows the relationships.

When a sine wave is applied to an RC parallel circuit, the total opposition to current flow is a combination of R and X_c. This total opposition is known as impedance, symbolized by Z. However, unlike series RC circuits, the combination of R and X_c, in parallel, is not a vector sum. Therefore, solve for the total circuit current and then use Ohm's law to find the equivalent circuit impedance:

To solve for IT, use individual branch currents:

$$Ir = \sqrt{IR^2 + Ic^2}$$

To solve for equivalent Z, use Ohm's law:

$$Z = V_T/I_T$$

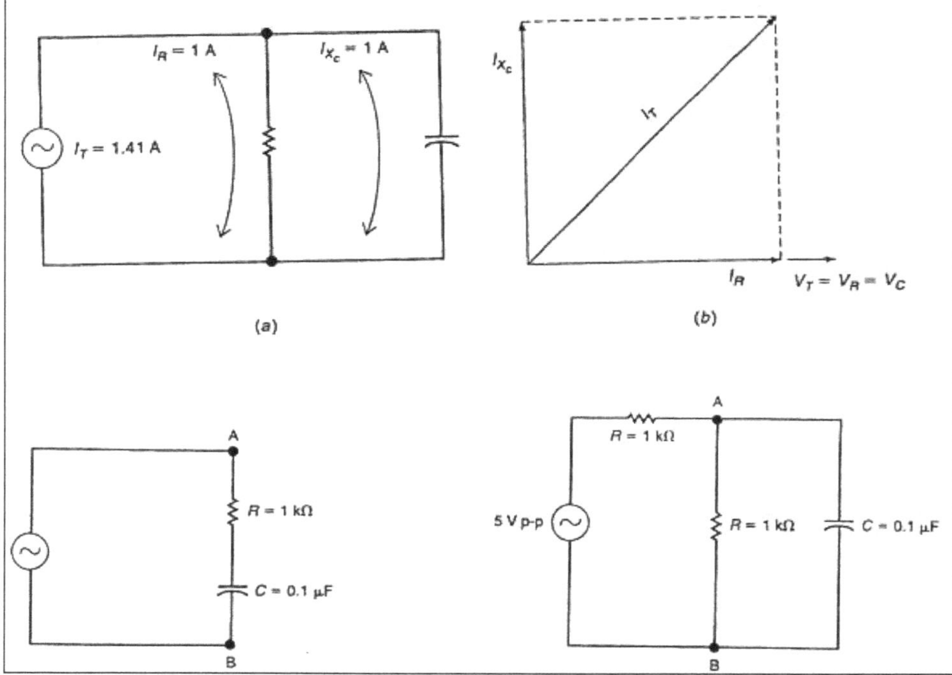

For parallel RC circuits, the impedance angle is based on the ratio between the branch currents and is not solved in a straightforward manner like series RC circuits. However, a parallel RC circuit can still be characterized as resistive or capacitive, but in a different manner. When R is 10 times greater than X_c the circuit is capacitive, because most of the current is flowing throughout the capacitive side. When X_c is 10 times greater than R, the circuit is resistive because most of the current is flowing through the resistor.

LC Circuit

An LC circuit is also called a tank circuit, tuned circuit or resonant circuit, is an electric circuit built with a capacitor denoted by the letter 'C' and an inductor denoted by the letter 'L' connected together. These circuits are used for producing signals at a particular frequency or accepting a signal from a more composite signal at a particular frequency. LC circuits are basic electronics components in various electronic devices, especially in radio equipment used in circuits like tuners, filters, frequency mixers and oscillators. The main function of an LC circuit is generally to oscillate with minimum damping.

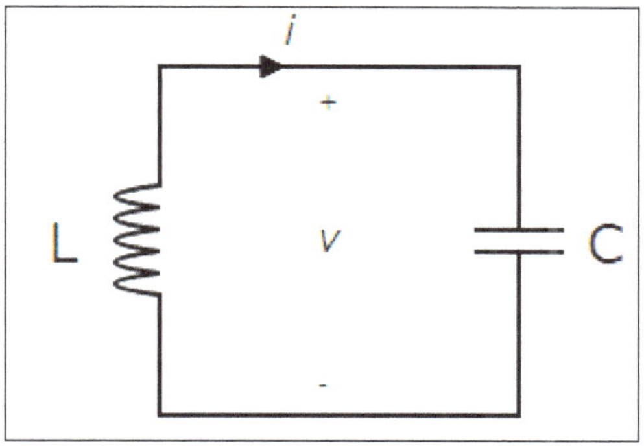

LC Circuit

Series LC Circuit Resonance

In the series LC circuit configuration, the capacitor 'C' and inductor 'L' both are connected in series that is shown in the following circuit. The sum of the voltage across the capacitor and inductor is simply the sum of the whole voltage across the open terminals. The flow of current in the +Ve terminal of the LC circuit is equal to the current through both the inductor (L) and the capacitor (C),

$$v = v_L + v_C$$

$$i = i_L = i_C$$

When the 'X_L' inductive reactance magnitude increases, then the frequency also increases. In the same way while 'X_C' capacitive reactance magnitude decreases, then the frequency decreases.

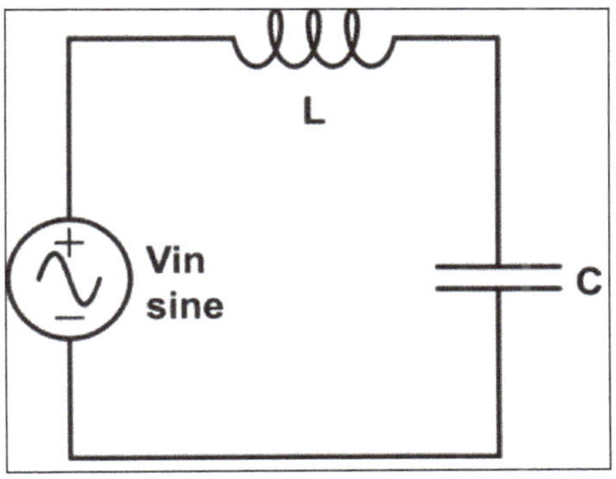

Series LC Circuit Resonance

At one specific frequency, the two reactances X_L and X_C are the same in magnitude but reverse in sign. So this frequency is called the resonant frequency which is denoted by for the LC circuit.

Therefore, at resonance,

$$X_L = -X_C$$

$$\omega_L = 1/\omega_C$$

$$\omega = \omega_o = 1/\sqrt{LC}$$

Which is termed as the resonant angular frequency of the circuit. Changing angular frequency into frequency, the following formula is used,

$$f_o = \omega_o/2\pi\sqrt{LC}$$

In a series resonance LC circuit configuration, the two resonances X_C and X_L cancel each other out. In actual, rather than ideal components, the flow of current is opposed, generally by the resistance of the windings of the coil. Therefore, the current supplied to the circuit is a max at resonance.

An acceptance circuit is defined as when the In the Lt, f_o is the maximum and the impedance of the circuit is minimized.

For $f<f_o$, $X_L << (-X_C)$. Thus, the circuit is capacitive.

For $f<f_o$, $X_L >> (-X_C)$. Thus, the circuit is inductive.

Parallel LC Circuit Resonance

In the parallel LC circuit configuration, the capacitor 'C' and inductor 'L' both are connected in parallel that is shown in the following circuit. The sum of the voltage across the capacitor and inductor is simply the sum of the whole voltage across the open terminals. The flow of current in the +Ve terminal of the LC circuit is equal to the current through both the inductor (L) and the capacitor (C).

$$v = v_L = v_C$$

$$i = i_L + i_C$$

Let the internal resistance 'R' of the coil. When two resonances X_C and XL, the reactive branch branch currents are the same and opposed. Therefore, they cancel out each other to give the smallest amount of current in the key line. When the total current is minimum in this state, then the total impedance is max. The resonant frequency is given by,

$$f_o = \omega_o/\ 2\pi = 1/2\pi\sqrt{LC}$$

The current of any reactive branch is not minimum at resonance, but each is given individually by separating source voltage 'V' by reactance 'Z'.

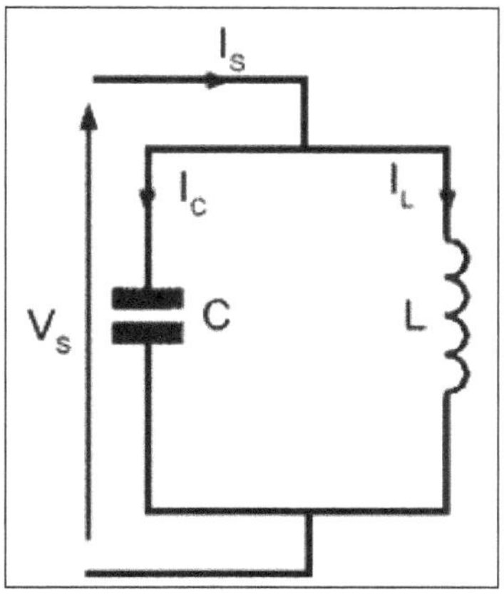

Parallel LC Circuit Resonance

Hence, according to Ohm's law I=V/Z

A rejector circuit can be defined as, when the line current is minimum and total impedance is max at f_o , circuit is inductive when below f_o and the circuit is capacitive when above f_o.

Applications of LC Circuit

- The applications of the resonance of the series and parallel LC circuits mainly involve in communications systems and signal processing.
- The common application of an LC circuit is, tuning radio TXs and RXs. For instance, when we tune a radio to an exact station, then the circuit will set at resonance for that specific carrier frequency.
- A series resonant LC circuit is used to provide voltage magnification.
- A parallel resonant LC circuit is used to provide current magnification and also used in the RF amplifier circuits as the load impedance, the amplifier's gain is max at the resonant frequency.
- Both series and parallel resonant LC circuits are used in induction heating.
- These circuits perform as electronic resonators, which are an essential component in various applications like amplifiers, oscillators, filters, tuners, mixers, graphic tablets, contactless cards and security tags X_L and X_C.

RL Circuit

A circuit with resistance and self-inductance is known as an *RL* circuit. Figure shows an *RL* circuit consisting of a resistor, an inductor, a constant source of emf, and switches S_1 and S_2. When S_1 is closed, the circuit is equivalent to a single-loop circuit consisting of a resistor and an inductor connected across a source of emf When S_1 is opened and S_2 is closed, the circuit becomes a single-loop circuit with only a resistor and an inductor.

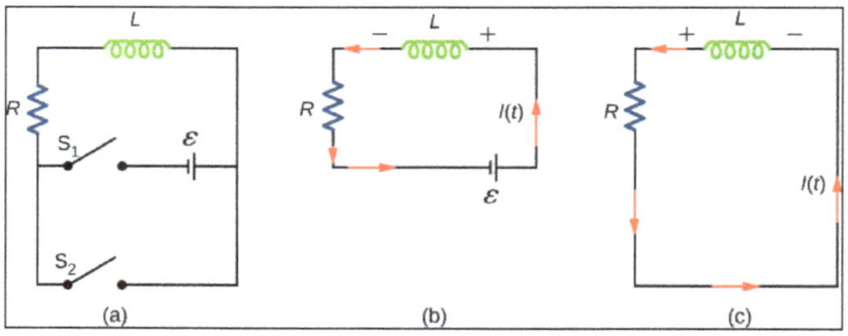

(a) An RL circuit with switches S_1 and S_2. (b) The equivalent circuit with S_1 closed and S_2 open. (c) The equivalent circuit after S_1 is opened and S_2 is closed.

We first consider the RL circuit of figure. Once S_1 is closed and S_2 is open, the source of emf produces a current in the circuit. If there were no self-inductance in the circuit, the current would rise immediately to a steady value of ϵ/R. However, from Faraday's law, the increasing current produces an emf $V_L=-L(dI/dt)$ across the inductor. In accordance with Lenz's law, the induced emf counteracts the increase in the current and is directed as shown in the figure. As a result, I(t) starts at zero and increases asymptotically to its final value.

Applying Kirchhoff's loop rule to this circuit, we obtain.

$$\epsilon - L\frac{dI}{dt} - IR = 0,$$

Which is a first-order differential equation for I(t). Its similarity to the equation for a capacitor and resistor in series. Similarly, the solution to Equation can be found by making substitutions in the equations relating the capacitor to the inductor. This gives, is the inductive time constant of the circuit.

$$I(t) = \frac{\epsilon}{R}\left(1 - e^{-Rt/L}\right)$$
$$= \frac{\epsilon}{R}\left(1 - e^{-t/\tau L}\right),$$

where,

$$\tau L = L/R$$

The current I(t) is plotted in Figure. It starts at zero, and as $t\to\infty$, I(t) approaches ϵ/R asymptotically. The induced emf VL(t) is directly proportional to dI/dt, or the slope of the curve. Hence, while at its greatest immediately after the switches are thrown, the induced emf decreases to zero with time as the current approaches its final value of ϵ/R. The circuit then becomes equivalent to a resistor connected across a source of emf.

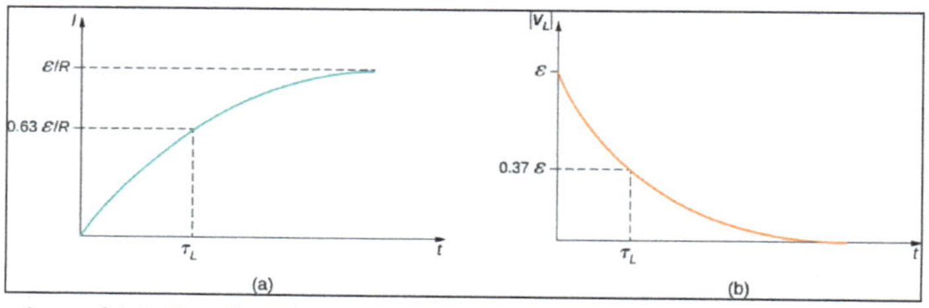

Time variation of (a) the electric current and (b) the magnitude of the induced voltage across the coil in the circuit of Figure.

The energy stored in the magnetic field of an inductor is,

$$UL = \frac{1}{2}LI^2.$$

Thus, as the current approaches the maximum current ϵ/R , the stored energy in the inductor increases from zero and asymptotically approaches a maximum of L(ϵ/R)²/2 .

The time constant τL tells us how rapidly the current increases to its final value.

At, t = τL, the current in the circuit is,

$$I(\tau L) = \frac{\epsilon}{R}(1 - e^{-1}) = 0.63\frac{\epsilon}{R}$$

Which is 63% of the final ϵ/R. The smaller the inductive time constant τL=L/R, the more rapidly the current approaches ϵ/R.

We can find the time dependence of the induced voltage across the inductor in this circuit by using VL (t)=−L(dI/dt).

$$V_L(t) = -L\frac{dI}{dt} = -\epsilon e^{-t/\tau L}$$

The magnitude of this function is plotted in figure. The greatest value of L(dI/dt) is ϵ; it occurs when dI/dt is greatest, which is immediately after S_1 is closed and S_2 is opened. In the approach to steady state, dI/dt decreases to zero. As a result, the voltage across the inductor also vanishes as $t \to \infty$.

The time constant τL also tells us how quickly the induced voltage decays. At t = τL the magnitude of the induced voltage is.

$$|VL(\tau L)| = \epsilon e^{-1} = 0.37\epsilon = 0.37\,V(0).$$

The voltage across the inductor therefore drops to about 37% of its initial value after one time constant. The shorter the time constant τL, the more rapidly the voltage decreases.

After enough time has elapsed so that the current has essentially reached its final value, the positions of the switches in Figure a are reversed, giving us the circuit in part (c). At t=0, the current in the circuit is I(0)=ϵ/R. With Kirchhoff's loop rule we obtain,

$$IR + L\frac{dI}{dt} = 0$$

The solution to this equation is similar to the solution of the equation for a discharging capacitor, with similar substitutions. The current at time t is then,

$$I(t) = \frac{\epsilon}{R}e^{-t/\tau L}.$$

The current starts at I(0)=ε/R and decreases with time as the energy stored in the inductor is depleted.

The time dependence of the voltage across the inductor can be determined from VL=−L (dI/dt):

$$VL(0) = \epsilon e^{-t/\tau L}.$$

This voltage is initially VL(0)=ε , and it decays to zero like the current. The energy stored in the magnetic field of the inductor, $LI^2/2$, also decreases exponentially with time, as it is dissipated by Joule heating in the resistance of the circuit.

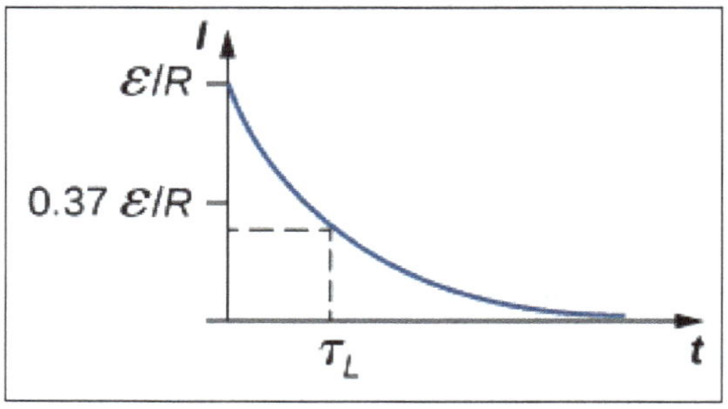

Figure: Time variation of electric current in the RL circuit. The induced voltage across the coil also decays exponentially.

RL Series Circuit

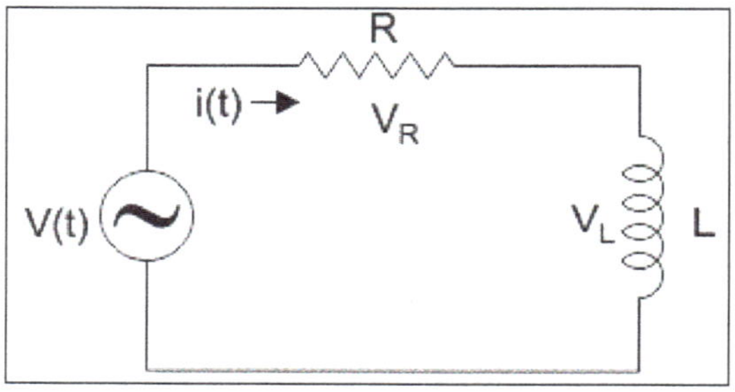

Consider a simple RL circuit in which resistor, R and inductor, L are connected in series with a voltage supply of V volts. Consider the current flowing in the circuit is I (amp) and current through resistor and inductor is I_R and I_L respectively. Since both resistance and inductor are connected in series, so the current in both the elements and the circuit remains the same. i.e $I_R = I_L = I$. Let V_R and V_l be the voltage drop across resistor and inductor.

Applying Kirchhoff voltage law (i.e sum of voltage drop must be equal to apply voltage) to this circuit we get,

$$V = V_R + V_L$$

Phasor Diagram for RL Circuit

Before drawing the phasor diagram of series RL circuit, one should know the relationship between voltage and current in case of resistor and inductor.

Resistor

In case of resistor, the voltage and the current are in same phase or we can say that the phase angle difference between voltage and current is zero.

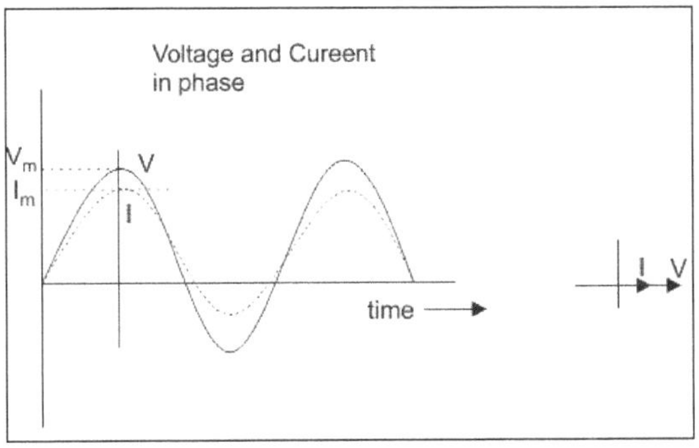

Inductor

In inductor, the voltage and the current are not in phase. The voltage leads that of current by 90° or in other words, voltage attains its maximum and zero value 90° before the current attains it.

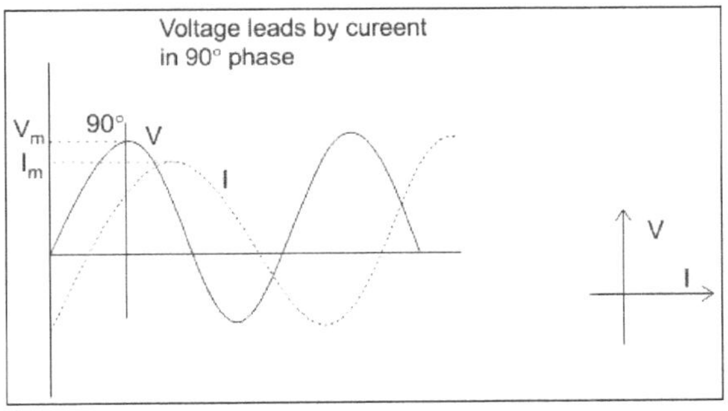

RL Circuit

For drawing the phasor diagram of series RL circuit; follow the following steps:

Step I. In case of series RL circuit, resistor and inductor are connected in series, so current flowing in both the elements are same i.e $I_R = I_L = I$. So, take current phasor as reference and draw it on horizontal axis as shown in diagram.

Step II. In case of resistor, both voltage and current are in same phase. So draw the voltage phasor, V_R along same axis or direction as that of current phasor. i.e V_R is in phase with I.

Step III. We know that in inductor, voltage leads current by 90°, so draw V_L (voltage drop across inductor) perpendicular to current phasor.

Step IV. Now we have two voltages V_R and V_L. Draw the resultant vector(V_G) of these two voltages. Such as,

$V_R^2+V_L^2=V_G^2$ and from right angle triangle we get, phase angle $\theta=\tan^{-1}(V_L^1=V_R^2)$

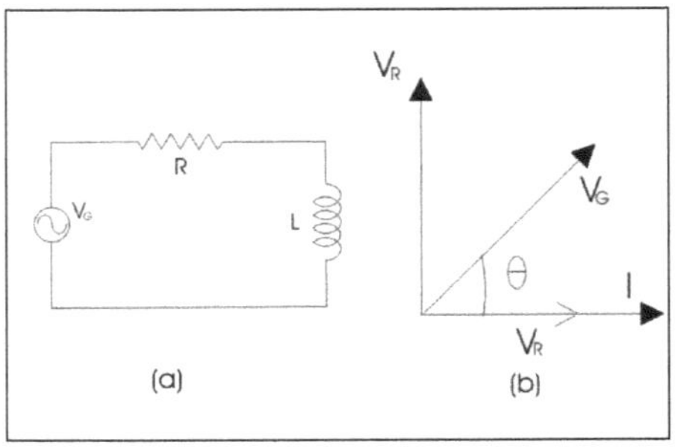

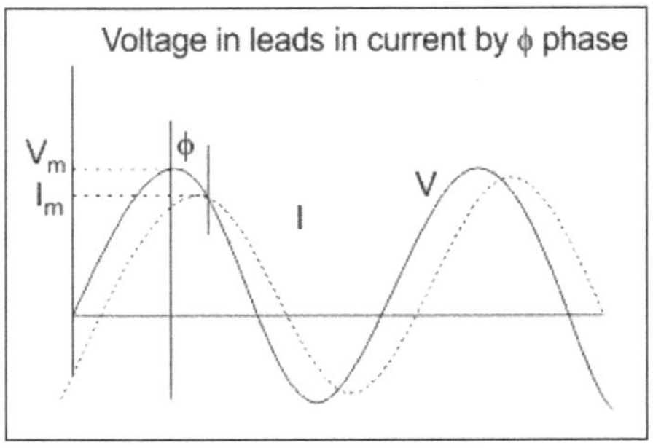

In case of pure resistive circuit, the phase angle between voltage and current is zero and in case of pure inductive circuit, phase angle is 90° but when we combine both resistance and inductor, the phase angle of a series RL circuit is between 0° to 90°.

Impedance of Series RL Circuit

The impedance of series RL circuit opposes the flow of alternating current. The impedance of series RL Circuit is nothing but the combine effect of resistance (R) and inductive reactance (X_L) of the circuit as a whole. The impedance Z in ohms is given by, $Z = (R^2 + X_L^2)^{0.5}$ and from right angle triangle, phase angle $\theta = \tan^{-1}(X_L/R)$.

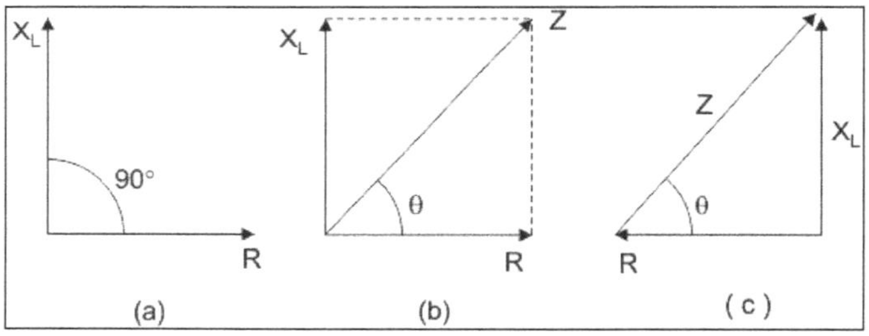

Series RL Circuit Analysis

In series RL circuit, the values of frequency f, voltage V, resistance R and inductance L are known and there is no instrument for directly measuring the value of inductive reactance and impedance; so, for complete analysis of series RL circuit, follow these simple steps:

Step 1. Since the value of frequency and inductor are known, so firstly calculate the value of inductive reactance X_L: $X_L = 2\pi fL$ ohms. Step 2. From the value of X_L and R, calculate the total impedance of the circuit which is given by,

$$Z = \sqrt{(R^2 + X_L^2)}$$

Step 2. Calculate the total phase angle for the circuit $\theta = \tan^{-1}(X_L / R)$.

Step 3. Use Ohm's Law and find the value of the total current: $I = V/Z$ amp. Step 5. Calculate the voltages across resistor R and inductor L by using Ohm's Law. Since the resistor and the inductor are connected in series, so current in them remains the same.

$$V_R = RI \text{ Volts and } V_L = X_L I \text{ Volts}$$

Power in RL Circuit

In series RL circuit, some energy is dissipated by the resistor and some energy is alternately stored and returned by the inductor-

1. The instantaneous power deliver by voltage source V is P = VI (watts)
2. Power dissipated by the resistor in the form of heat, $P = I^2R$ (watts)
3. The rate at which energy is stored in inductor

$$P = V_L I = LI\frac{dI}{dt}\ (watts)$$

So, total power in series RL circuit is given by adding the power dissipated by the resistor and the power absorbed by the inductor.

$$P = I^2R + LI\frac{dI}{dt}$$

Power triangle for series RL circuit is shown below,

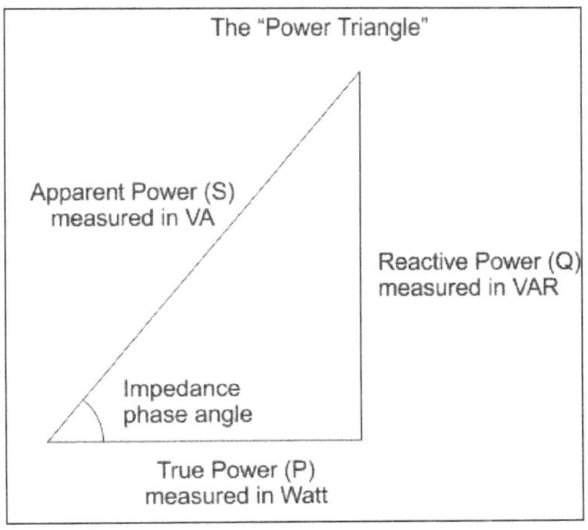

The electrical power factor cos θ is defined as ratio of the true power to apparent power.

Variation of Impedance and Phase Angle with Frequency

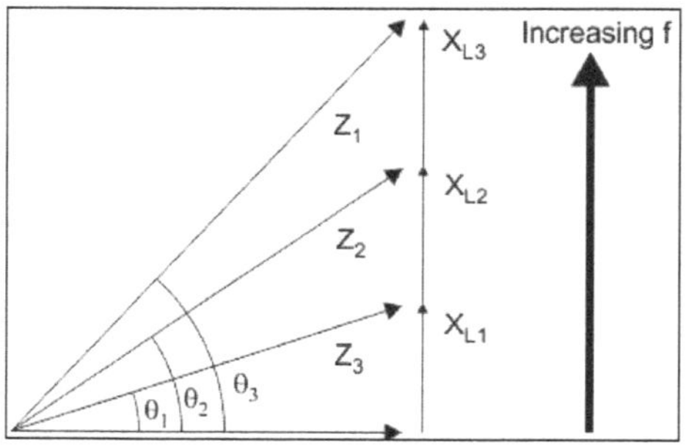

The above diagram shows the impedance triangle. The base of this impedance triangle represents resistance. The resistance is independent of frequency; so, if frequency increases or decreases, resistance remains constant. The formula for inductive reactance is $X_L = 2\pi fL$. So, if frequency increases inductive reactance X_L also increases and if inductive reactance increases, total impedance of circuit also increases and this leads to variation in phase angle θ with frequency. So, in series RL circuit if frequency increases,

1. Inductive reactance also increases as it is directly proportional to frequency.
2. Total impedance Z increases.
3. Phase angle θ increases.
4. Resistance remains constant.

Expression for Current Flowing in Series RL Circuit

Consider a circuit in which resistance is connected in series with inductor and voltage source of V volts, is applied across it. Initially the switch is open. Let us say at time 't' we close the switch and the current 'i' starts flowing in the circuit but it does not attains its maximum value rapidly due to the presence of inductor in the circuit as we know inductor has a property to oppose the change in the current flowing through it.

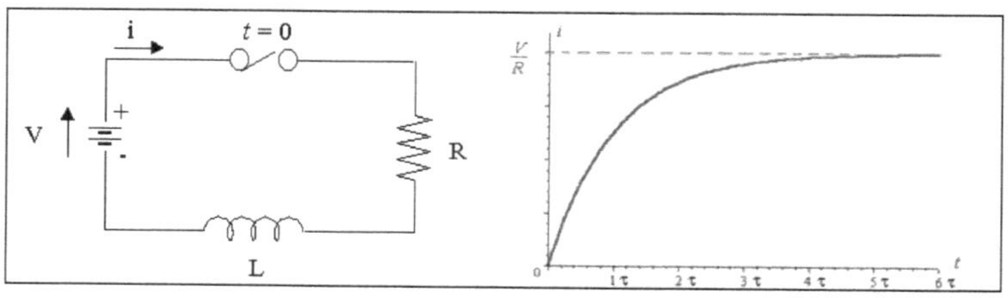

Apply Kirchhoff's voltage law in the above series RL circuit,

$$V - iR - L\frac{di}{dt} = 0$$

Rearranging the above equation,

$$V - iR = L\frac{di}{dt}$$

$$\frac{dt}{L} = \frac{di}{v - iR}$$

Integrating both sides, we get,

$$\int_0^t \frac{dt}{L} = \int_0^i \frac{di}{V - iR}$$

$$or \frac{t}{L} = \int_0^i \frac{di}{V - iR}$$

Now integrate right hand side by using substitution method,

$$Let\, z = V - iR\, or \frac{dz}{di} = -R$$

Substituting the values we get,

$$or \frac{t}{L} = -\frac{1}{R}\int \frac{dz}{z}$$

We know that integration of,

$$\frac{1}{z} = In(z)$$

So we get,

$$-\frac{Rt}{L} = In(V - iR)_o^i$$

By applying limits we get,

$$-\frac{Rt}{L} = In(V - iR) - In(V)$$

Simplifying again,

$$-\frac{Rt}{L} = In\frac{V - iR}{V}$$

Taking antilog on both sides,

$$e^{-\frac{RT}{L}} = e^{in}\,\frac{v - iR}{v}$$

We know that $e^{\ln} = x$, so we get,

$$e^{-\frac{RT}{L}} = \frac{v - iR}{V}$$

Moving the term containing 'i' on one side we get,

$$i = \frac{V(1 - e^{-\frac{RT}{L}})}{R}$$

The term L/R in the equation is called the Time Constant, (τ) of the RL series circuit, and it is defined as time taken by the current to reach its maximum steady state value and the term V/R represents the final steady state value of current in the circuit.

RL Parallel Circuit

In RL parallel circuit resistor and inductor are connected in parallel with each other and this combination is supplied by a voltage source, V_{in}. The output voltage of circuit is V_{out}. Since the resistor and inductor are connected in parallel, the input voltage is equal to output voltage but the currents flowing in resistor and inductor are different.

The parallel RL circuit is not used as filter for voltages because in this circuit, the output voltage is equal to input voltage and for this reason it is not commonly used as compared to series RL circuit.

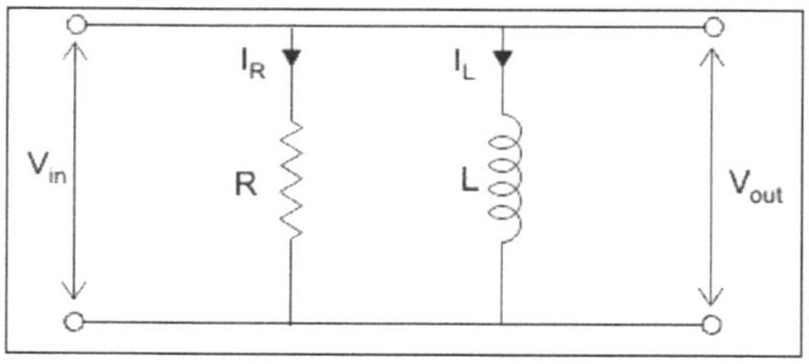

Let us say: I_T = the total current flowing from voltage source in amperes.

I_R = the current flowing in the resistor branch in amperes.

I_L = the current flowing in the inductor branch in amperes.

θ = angle between I_R and I_T.

So the total current I_T,

$$I_T^2 = I_R^2 + I_L^2$$

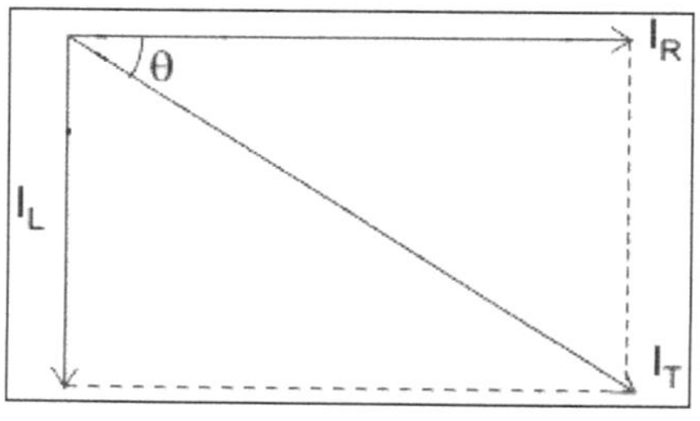

In complex form the currents are written as,

$$I_R = \frac{V_{in}}{R}$$

$$I_R = \frac{V_{in}}{j\omega L}$$

$$\Rightarrow I_L = -j\frac{V_{in}}{\omega L}\left(Where \frac{1}{j} = -j\right)$$

$$There\,f\,ore\;totel\;current\;I_T = \frac{V_{in}}{R - \frac{V_{in}}{j\omega L}}.$$

Impedance of Parallel RL Circuit

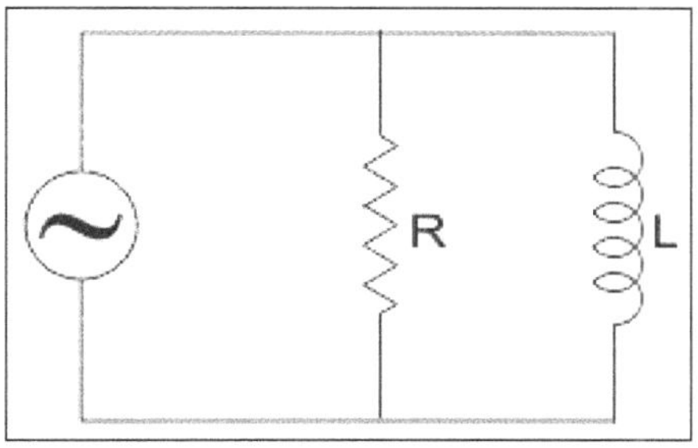

Let, Z = total impedance of the circuit in ohms

R = resistance of circuit in ohms

L = inductor of circuit in Henry

X_L = inductive reactance in ohms

Since resistance and inductor are connected in parallel, the total impedance of the circuit is given by,

$$\frac{1}{Z} = \frac{1}{R} + \frac{1}{jXL}$$

$$Z = \frac{R*({}_jX_L)}{R + {}_jX_L}$$

In order to remove "j" from the denominator multiply and divide numerator and denominator by $(R - jX_L)$

$$z = \frac{R*(jX_L)}{R + jX_L} * \frac{R - jX_L}{R - jX_L}$$

$$= \frac{(jRX_L)(R - jX_L)}{(R + jX_L)(R - jX_L)} = \frac{jR^2X_L - J^2R^2X_L^2}{R_2 + X_L^2}$$

$$(Since\, J^2 = -1)$$

$$= \frac{RX_L^2 + jR^2\ X_L}{R^2 + X_L^2}$$

$$Z = \frac{RX_L^2}{R^2 + X_L^2} + j\frac{R^2 X_L}{R^2 + X_L^2}$$

Analysis of a Parallel RL Circuit

In parallel RL circuit, the values of resistance, inductance, frequency and supply voltage are known for finding the other parameters of RL parallel circuit follow these steps:

Step 1. Since the value of frequency is already known, we can easily find the value of inductive reactance X_L

$$X_L = 2\pi f\, L$$

Step 2. We know that in parallel circuit, the voltage across inductor and resistor remains the same so,

$$V_R V_L = V$$

Step 3. Use Ohm's law to find the current flowing through inductor and resistor,

$$I_R = \frac{V}{R} and\ I_L = \frac{V}{XL}$$

Step 4. Now calculate the total current,

$$I_T^2 = I_R^2 + I_L^2$$

Step 5. Determine the phase angles for resistor and inductor and for parallel circuit, its always

$$\theta_R = 0^\circ\ and\ \theta_L = -90^\circ$$

Step 6. Since we have already calculated the total current flowing in the circuit and voltage V, by using Ohm's law; we can easily calculate the total impedance:

$$Z = \frac{V}{I_T}$$

Step 7. Now calculate the total phase angle for the circuit which is given by,

$$\theta_T = \tan^{-1}\frac{IL}{IR}$$

The total phase angle of a parallel RL circuit always lies between 0° to -90°. It is 0° for pure resistive circuit and -90° for pure inductive circuit.

RLC Circuit

The RLC circuit is the electrical circuit consisting of a resistor of resistance R, a coil of inductance L, a capacitor of capacitance C and a voltage source arranged in series.

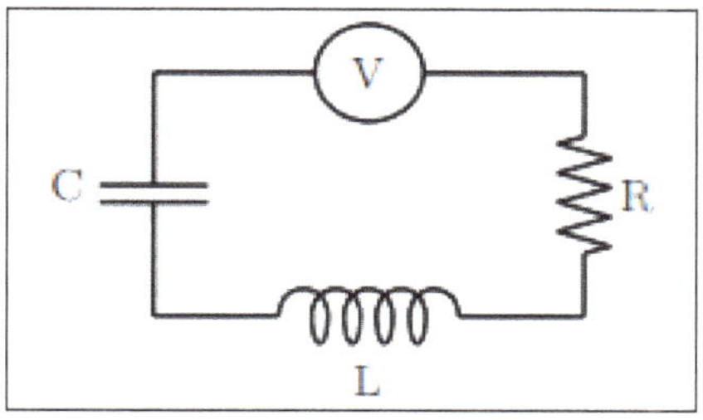

If the charge on the capacitor is Q and the current flowing in the circuit is I, the voltage across R, L RI, $L\frac{dI}{dt}$ and $\frac{Q}{C}$ respectively. By the Kirchhoff's law that says that the voltage between any two points has to be independent of the path used to travel between the two points.

$$LI'(t)+RI(t)+\frac{1}{C}Q(t)=V(t)$$

Assuming that R, L, C and V are known, this is still one differential equation in two unknowns, I and Q. However the two unknowns are related by $I(t)=\frac{dQ}{dt}(t)$ so that

$$LQ''(t) + RI'(Q)(t)+\frac{1}{C}Q(t)\ V(t)$$

or, differentiating with respect to t and then subbing in $\frac{dQ}{dt}(t) = I(t)$,

$$LI''(t) + RI'(t) + \frac{1}{C}I(t)\ VI'(t)$$

For an ac voltage source, choosing the origin of time so that V (0) = 0, V (t) = Eo sin (ωt) and the differential equation becomes.

$$LI''(t) + RI'(t) + \frac{1}{C}I(t) = \omega E0\ \cos(\omega t)$$

The General Solution:

We first guess one solution of by trying I_p (t) = A sin (ωt – φ) with the amplitude A and phase ϕ to be determined. That is, we are guessing that the circuit responds to an oscillating applied voltage with a current that oscillates with the same rate. For I_p (t) to be a solution, we need.

$$LI_p''(t)+ RI_p'(t) + \frac{1}{C}I_p(t)= \omega E_0\ \cos(\omega t)$$

$$-L\omega^2 A\sin(\omega t - \varphi) + R\omega A\cos(\omega t - \varphi) + \frac{1}{C}A\sin(\omega t - \varphi) = \omega E0 \cos(\omega t)$$
$$= \omega E0 \cos(\omega t - \varphi + \varphi)$$

and hence.

$$\left(\frac{1}{C} - L\omega^2 \text{ Asin}(\omega\text{t-}\varphi)\right) + \text{R}\omega\text{Acos}(\omega\text{t-}\varphi) = \omega\text{E}_0\cos(\varphi)\cos(\omega\text{t-}\varphi) - \omega\text{E}_0 \sin(\varphi)\sin(\omega\text{t-}\varphi)$$

Matching coefficients of sin (ωt−ϕ) and cos (ωt−ϕ) on the left and right hand sides gives.

$$\left(\text{L}\omega^2 - \frac{1}{\text{C}}\right)\text{A} = \omega\text{E}_0 \sin(\varphi)$$
$$\text{R}\omega\text{A} = \omega\text{E}_0 \cos(\varphi)$$

It is now easy to solve for A and ϕ

$$\tan(\varphi) = \frac{L\omega^2 - \frac{1}{c}}{R\omega} \qquad \Rightarrow \varphi = \tan^{-1}\left(\frac{\text{L}\omega}{R} - \frac{1}{\text{RC}\omega}\right)$$

$$\Rightarrow \sqrt{(\text{L}\omega^2 - \frac{1}{\text{C}})^2 + \text{R}^2\omega^2}\text{A} = \omega\text{E}_0 \Rightarrow \text{A} = \frac{\omega\text{E}_0}{\sqrt{(\text{L}\omega^2 - \frac{1}{\text{C}})^2 + \text{R}^2\omega^2}}$$

Naturally, different input frequencies ω give different output amplitudes A. Here is a graph of A against ω, with all other parameters held fixed.

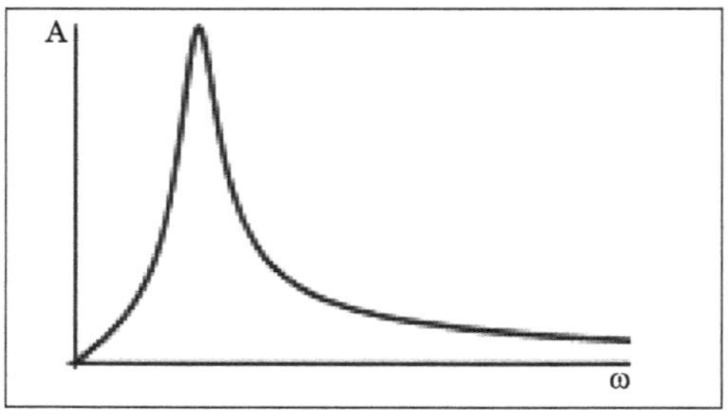

There is a small range of frequencies that give a large amplitude response. This is the phenomenon of resonance. It has been dramatically illustrated in, for example, the collapse of the Tacoma narrows bridge.

Subtracting $LI_p''(t) + RI_p'(t) + \frac{1}{C}I_p(t) = \omega E_0 \cos(\omega t)$ from gives

$$LI''(t) + RI'(t) + \frac{1}{C}I(t) = \omega E0 \cos(\omega t)$$

$$L(I - I_p)''(t) + R(I - I_p)'(t) + \frac{1}{c}(1-1_p)(t) = 0$$

That is, any solution of differs from $I_p(t)$ by a solution of,

$$LI''(t) + RI'(t) + \frac{1}{C}I(t) = 0$$

This is called the complementary homogeneous equation for. We now guess many solutions to (1_c) by trying I (t) = e^{rt} with the constant r to the determined. This guess is a solution of (1_c) if and only if

$$Lr^2e^{rt} + Rre^{rt} + \frac{1}{c}e^{rt} = 0 \Leftrightarrow Lr^2 + \mathrm{R}r + \frac{1}{c} = 0 \Leftrightarrow r = \frac{-R \pm \sqrt{R^2 - 4L/c}}{2L} \equiv r_{1,2}$$

We now know that e^{r1t} and e^{r2t} both obey (1_c). Because (1_c) is linear and homogeneous, this forces $c_1e^{r1t} + c_2e^{r2t}$ to also be a solution, for any values of the constants c_1 and c_2. (To check this, just sub $c_1e^{r1t} + c_2e^{r2t}$ into (1_c).) Assuming that $R^2 \neq 4L/C$, r_1 and r_2 are different and the general solution to (1_c) is c_1e^{r1t} + c2e^{r2t}. (It is reasonable to guess that, to solve a differential equation involving a second derivative, one has to integrate twice so that the general solution contains two arbitrary constants.) Then, the general solution of is,

$$I(t) = c1e^{r1t} + c2e^{r2t} + A\sin(\omega t - \varphi)$$

With r_1, r_2 given in and A, ϕ given in. The arbitrary constants c_1 and c_2 are determined by initial conditions. However, when e^{r1t} and e^{r2t} damp out quickly, as is often the case, their values are not very important.

Series RLC Circuit

Series RLC circuits consist of a resistance, a capacitance and an inductance connected in series across an alternating supply.

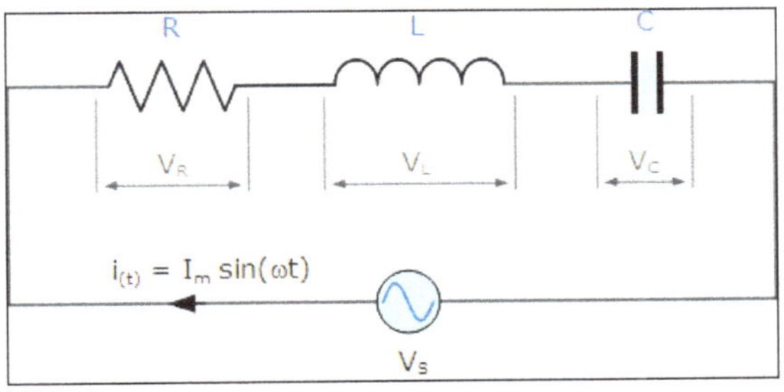

The series RLC circuit above has a single loop with the instantaneous current flowing through the loop being the same for each circuit element. Since the inductive and capacitive reactance's X_L and X_C are a function of the supply frequency, the sinusoidal response of a series RLC circuit will therefore vary with frequency, *f*. Then the individual voltage drops across each circuit element of R, L and C element will be "out-of-phase" with each other as defined by:

- $i_{(t)} = I_{max} \sin(\omega t)$

- The instantaneous voltage across a pure resistor, V_R is "in-phase" with current
- The instantaneous voltage across a pure inductor, V_L "leads" the current by 90°
- The instantaneous voltage across a pure capacitor, V_C "lags" the current by 90°
- Therefore, V_L and V_C are 180° "out-of-phase" and in opposition to each other.

For the series RLC circuit above, this can be shown as:

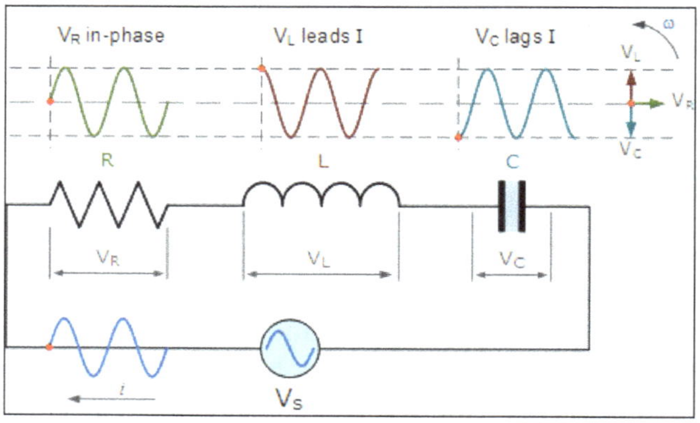

The amplitude of the source voltage across all three components in a series RLC circuit is made up of the three individual component voltages, V_R, V_L and V_C with the current common to all three components. The vector diagrams will therefore have the current vector as their reference with the three voltage vectors being plotted with respect to this reference as shown below.

Individual Voltage Vectors

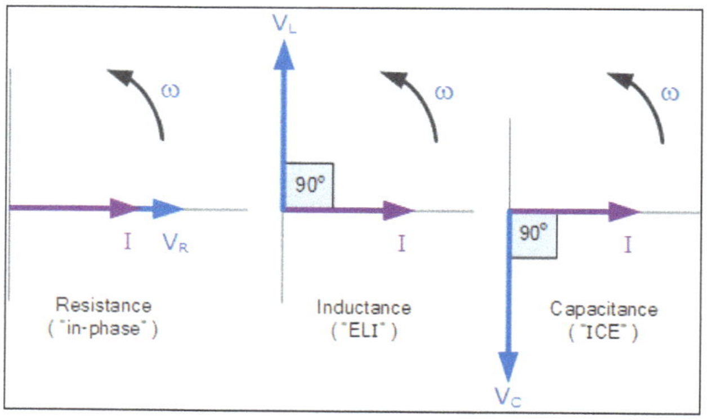

This means then that we cannot simply add together V_R, V_L and V_C to find the supply voltage, V_S across all three components as all three voltage vectors point in different directions with regards to the current vector. Therefore we will have to find the supply voltage, V_S as the Phasor Sum of the three component voltages combined together vectorially.

Kirchhoff's voltage law (KVL) for both loop and nodal circuits states that around any closed loop the sum of voltage drops around the loop equals the sum of the EMF's. Then applying this law to these three voltages will give us the amplitude of the source voltage, V_S.

Instantaneous Voltages for a Series RLC Circuit

$$\text{KVL: } V_s - V_R - V_L - V_c = 0$$

$$V_s - IR - L\frac{di}{dt} - \frac{Q}{C} = 0$$

$$V_s - IR - L\frac{di}{dt} + \frac{Q}{C}$$

The phasor diagram for a series RLC circuit is produced by combining together the three individual phasors above and adding these voltages vectorially. Since the current flowing through the circuit is common to all three circuit elements we can use this as the reference vector with the three voltage vectors drawn relative to this at their corresponding angles.

The resulting vector V_S is obtained by adding together two of the vectors, V_L and V_C and then adding this sum to the remaining vector V_R. The resulting angle obtained between V_S and i will be the circuits phase angle.

Phasor Diagram for a Series RLC Circuit

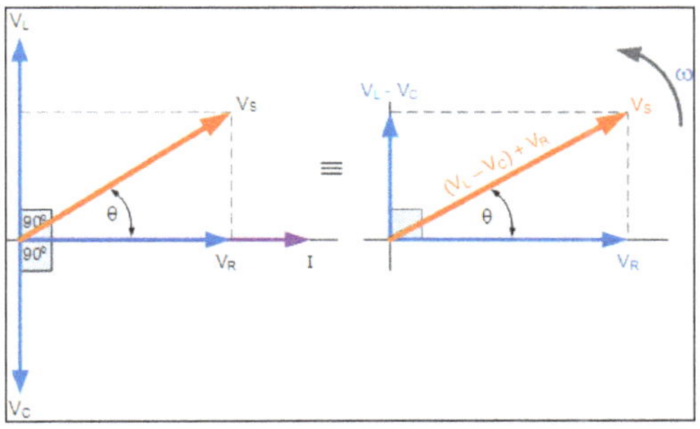

We can see from the phasor diagram on the right hand side above that the voltage vectors produce a rectangular triangle, comprising of hypotenuse V_S, horizontal axis V_R and vertical axis $V_L - V_C$ Hopefully you will notice then, that this forms our old favourite the Voltage Triangle and we can therefore use Pythagoras's theorem on this voltage triangle to mathematically obtain the value of V_S as shown.

Voltage Triangle for a Series RLC Circuit

$$V_s^2 = V_R^2 + (V_L - V_c)^2$$

$$V_s = \sqrt{V_R^2 + V_L - V_C)^2}$$

When using the above equation, the final reactive voltage must always be positive in value, that is the smallest voltage must always be taken away from the largest voltage we cannot have a negative voltage added to V_R so it is correct to have $V_L - V_C$ or $V_C - V_L$. The smallest value from the largest otherwise the calculation of V_S will be incorrect.

We know from above that the current has the same amplitude and phase in all the components of a series RLC circuit. Then the voltage across each component can also be described mathematically according to the current flowing through and the voltage across each element as,

$$V_R = iR\sin(\omega t + 0°) = i.R$$

$$V_L = iX_L \sin(\omega t + 90°) = i.jwL$$

$$V_C = iX_c \sin(\omega t - 90°) = i \cdot \frac{1}{jwc}$$

By substituting these values into the Pythagoras equation above for the voltage triangle will give us:

$$V_R = I.R,\ V_L = I.X_L,\ V_c = I.X_c$$

$$V_s = \sqrt{(I.R)^2 + (I.X_L - I.X_c)^2}$$

$$V_s = I.\sqrt{R^2 + (X_L - X_c)^2}$$

$$V_S 1 \times Z \quad \text{Where}: Z = \sqrt{R^2 + (X_L - X_c)^2}$$

So we can see that the amplitude of the source voltage is proportional to the amplitude of the current flowing through the circuit. This proportionality constant is called the Impedance of the circuit which ultimately depends upon the resistance and the inductive and capacitive reactances.

Then in the series RLC circuit above, it can be seen that the opposition to current flow is made up of three components, X_L, X_C and R with the reactance, X_T of any series RLC circuit being defined as: $X_T = X_L - X_C$ or $X_T = X_C - X_L$ whichever is greater. Thus the total impedance of the circuit being thought of as the voltage source required to drive a current through it.

The Impedance of a Series RLC Circuit

As the three vector voltages are out-of-phase with each other, X_L, X_C and R must also be "out-of-phase" with each other with the relationship between R, X_L and X_C being the vector sum of these three components. This will give us the RLC circuits overall impedance, Z. These circuit impedance's can be drawn and represented by an Impedance Triangle as shown below.

The Impedance Triangle for a Series RLC Circuit

$$Z^2 = R^2 + (X_L - X_c)^2$$

$$X_T = X_L - X_c$$

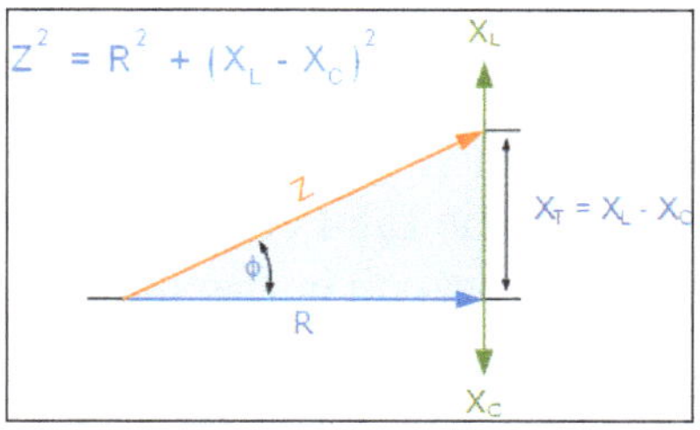

The impedance Z of a series RLC circuit depends upon the angular frequency, ω as do X_L and X_C If the capacitive reactance is greater than the inductive reactance, $X_C > X_L$ then the overall circuit reactance is capacitive giving a leading phase angle.

If the inductive reactance is greater than the capacitive reactance, $X_L > X_C$ then the overall circuit reactance is inductive giving the series circuit a lagging phase angle. If the two reactance's are the same and $X_L = X_C$ then the angular frequency at which this occurs is called the resonant frequency and produces the effect of resonance hopefully you will notice then, that (,).

Then the magnitude of the current depends upon the frequency applied to the series RLC circuit. When impedance, Z is at its maximum, the current is a minimum and, when Z is at its minimum, the current is at maximum. So the above equation for impedance can be re-written as:

$$\text{Impedance, } Z = \sqrt{R^2 + \left(\omega L - \frac{1}{\omega C}\right)^2}$$

The phase angle, θ between the source voltage, V_S and the current, i is the same as for the angle between Z and R in the impedance triangle. This phase angle may be positive or negative in value depending on whether the source voltage leads or lags the circuit current and can be calculated mathematically from the ohmic values of the impedance triangle as:

$$\cos\phi \frac{R}{Z} = \sin\phi = \frac{X_L - Z_c}{Z} \tan\phi = \frac{X_L - X_C}{R}$$

Parallel RLC Circuit

The Parallel RLC Circuit is the exact opposite to the series circuit.

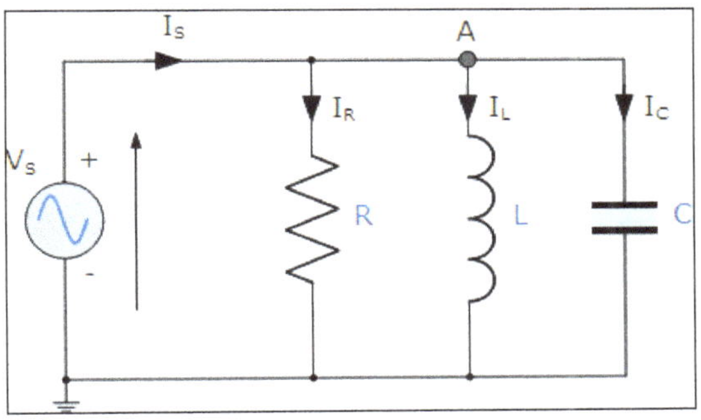

In the above parallel RLC circuit, we can see that the supply voltage, V_S is common to all three components whilst the supply current I_S consists of three parts. The current flowing through the resistor, I_R, the current flowing through the inductor, I_L and the current through the capacitor, I_C.

But the current flowing through each branch and therefore each component will be different to each other and to the supply current, I_S. The total current drawn from the supply will not be the mathematical sum of the three individual branch currents but their vector sum.

We can solve this circuit using the phasor or vector method but this time the vector diagram will have the voltage as its reference with the three current vectors plotted with respect to the voltage. The phasor diagram for a parallel RLC circuit is produced by combining together the three individual phasors for each component and adding the currents vectorially.

Since the voltage across the circuit is common to all three circuit elements we can use this as the reference vector with the three current vectors drawn relative to this at their corresponding angles. The resulting vector I_S is obtained by adding together two of the vectors, I_L and I_C and then adding this sum to the remaining vector I_R. The resulting angle obtained between V and I_S will be the circuits phase angle as shown below.

Phasor Diagram for a Parallel RLC Circuit

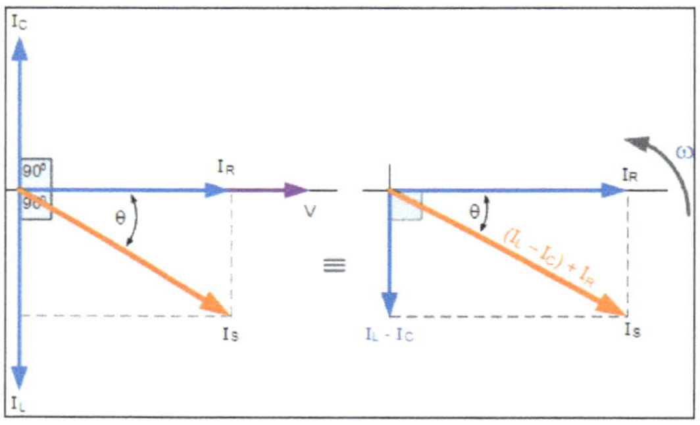

We can see from the phasor diagram on the right hand side above that the current vectors produce a rectangular triangle, comprising of hypotenuse I_S, horizontal axis I_R and vertical axis $I_L - I_C$ Hopefully you will notice then, that this forms a Current Triangle and we can therefore use Pythagoras's theorem on this current triangle to mathematically obtain the magnitude of the branch currents along the x-axis and y-axis and then determine the total current I_S of these components as shown.

Current Triangle for a Parallel RLC Circuit

$$I_S^2 = I_R^2 + (I_L - I_C)^2$$

$$I_S = \sqrt{I_R^2 + (I_L - I_C)^2}$$

$$\therefore I_S = \sqrt{\left(\frac{V^2}{R}\right) + \left(\frac{V}{X_L} - \frac{V}{X_C}\right)^2} = \frac{V}{Z}$$

$$\text{Where: } I_R = \frac{V}{R},\ I_L = \frac{V}{X_L},\ I_C = \frac{V}{X_C}$$

Since the voltage across the circuit is common to all three circuit elements, the current through each branch can be found using Kirchoff's Current Law, (KCL). Kirchoff's current law or junction law states that "the total current entering a junction or node is exactly equal to the current leaving that node", so the currents entering and leaving node "A" above are given as:

$$\text{KCL: } I_s - I_R - I_L - I_C = 0$$

$$I_S \frac{V}{R} - \frac{1}{L} \int V dt - C\frac{dv}{dt} = 0.$$

Taking the derivative, dividing through the above equation by C and rearranging gives us the following Second-order equation for the circuit current. It becomes a second-order equation because there are two reactive elements in the circuit, the inductor and the capacitor.

$$I_S \frac{d^2V}{dt^2} - \frac{dV}{RCdt} - \frac{V}{LC} = 0$$

$$\therefore I_{S(t)} - \frac{d^2V}{dt2} + \frac{dV}{dt}\frac{1}{RC} + \frac{1}{LC}V.$$

The opposition to current flow in this type of AC circuit is made up of three components: $X_L X_C$ and R with the combination of these three values giving the circuits impedance, Z. We know from above that the voltage has the same amplitude and phase in all the components of a parallel RLC circuit. Then the impedance across each component can also be described mathematically according to the current flowing through, and the voltage across each element as.

Impedance of a Parallel RLC Circuit

$$R = \frac{V}{I_R} \quad X_L = \frac{V}{I_L} \quad X_C = \frac{V}{I_C}$$

$$Z = \frac{1}{\sqrt{\left(\frac{1}{R}\right)^2 + \left(\frac{1}{X_L} - \frac{1}{X_C}\right)^2}}$$

$$\therefore \frac{1}{Z} = \frac{1}{\sqrt{\left(\frac{1}{R}\right)^2 + \left(\frac{1}{X_L} - \frac{1}{X_C}\right)^2}}$$

The final equation for a parallel RLC circuit produces complex impedance's for each parallel branch as each element becomes the reciprocal of impedance, (1/Z) with the reciprocal of impedance being called Admittance.

In parallel AC circuits it is more convenient to use admittance, symbol (Y) to solve complex branch impedance's especially when two or more parallel branch impedance's are involved (helps with the math's). The total admittance of the circuit can simply be found by the addition of the parallel admittances. Then the total impedance, Z_T of the circuit will therefore be $1/Y_T$ Siemens as shown.

Admittance of a Parallel RLC Circuit

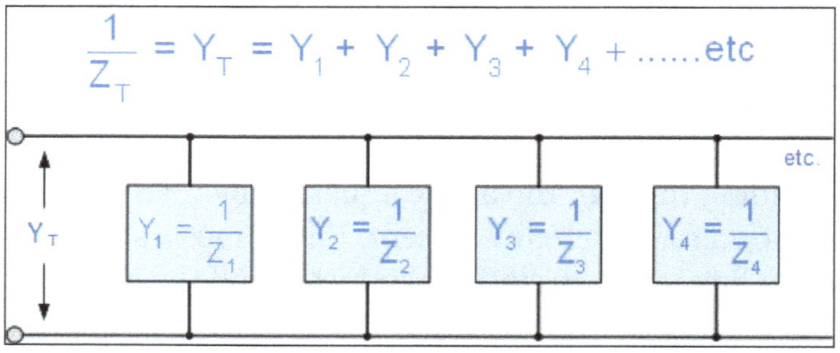

The new unit for admittance is the *Siemens*, abbreviated as S, (old unit mho's ℧, ohm's in reverse). Admittances are added together in parallel branches, whereas impedance's are added together in series branches. But if we can have a reciprocal of impedance, we can also have a reciprocal of resistance and reactance as impedance consists of two components, Rand X. Then the reciprocal of resistance is called Conductance and the reciprocal of reactance is called Susceptance.

Conductance, Admittance and Susceptance

The units used for conductance, admittance and susceptance are all the same namely Siemens (S), which can also be thought of as the reciprocal of Ohms or ohm^{-1}, but the symbol used for each element is different and in a pure component this is given as:

Admittance (Y)

Admittance is the reciprocal of impedance, Z and is given the symbol Y. In AC circuits admittance is defined as the ease at which a circuit composed of resistances and reactances allows current to flow when a voltage is applied taking into account the phase difference between the voltage and the current.

The admittance of a parallel circuit is the ratio of phasor current to phasor voltage with the angle of the admittance being the negative to that of impedance.

$$Y = \frac{1}{Z}\ [S]$$

Conductance (G)

Conductance is the reciprocal of resistance, R and is given the symbol G. Conductance is defined as the ease at which a resistor (or a set of resistors) allows current to flow when a voltage, either AC or DC is applied.

$$G = \frac{1}{R}[S]$$

Susceptance (B)

Susceptance is the reciprocal of of a pure reactance, X and is given the symbol B. In AC circuits susceptance is defined as the ease at which a reactance (or a set of reactances) allows an alternating current to flow when a voltage of a given frequency is applied.

Susceptance has the opposite sign to reactance so Capacitive susceptance B_C is positive, (+ve) in value while Inductive susceptance B_L is negative, (-ve) in value.

$$B_L = \frac{1}{X_L}\ [S]$$

$$B_c = \frac{1}{X_C}\ [S]$$

We can therefore define inductive and capacitive susceptance as being:

$$B_L = B\angle -90^\circ - jB \text{ and } B_C = B\angle +90^\circ = 0 + jB$$

In AC series circuits the opposition to current flow is impedance, Z which has two components, resistance R and reactance, X and from these two components we can construct an impedance triangle. Similarly, in a parallel RLC circuit, admittance, Y also has two components, conductance, G and susceptance, B. This makes it possible to construct an admittance triangle that has a horizontal conductance axis, G and a vertical susceptance axis, jB as shown.

Admittance Triangle for a Parallel RLC Circuit

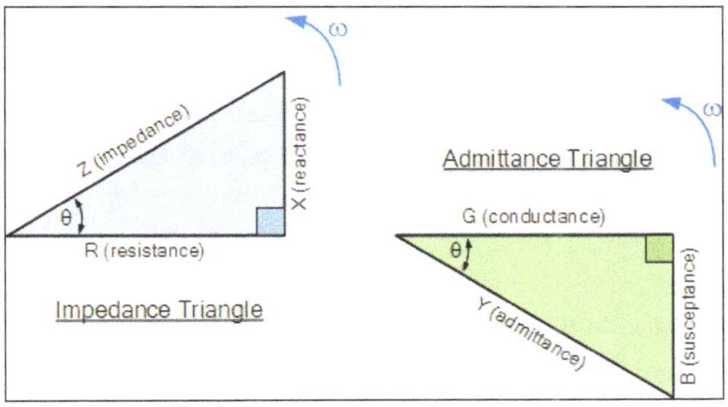

We have an admittance triangle, we can use Pythagoras to calculate the magnitudes of all three sides as well as the phase angle as shown.

From Pythagoras,

$$Y = \sqrt{G^2 + (B_L - B_c)^2}$$

$$\text{where: } Y=\frac{1}{Z} \quad G=\frac{1}{R}$$

$$B_L=\frac{1}{\omega L} \quad B_C=\omega C$$

Then we can define both the admittance of the circuit and the impedance with respect to admittance as:

$$\text{Admittance: } Y=\sqrt{\left(\frac{I}{R}\right)^2+\left(\frac{1}{(\omega L)}(\omega c)\right)^2}$$

$$\text{Impedance: } Z=\frac{1}{Y}=\frac{1}{\sqrt{\left(\frac{I}{R}\right)^2+\left(\frac{1}{(\omega L)}(\omega c)\right)^2}}$$

Giving us a power factor angle of:

$$\cos\phi\ \frac{G}{Y} \qquad \phi=\cos^{-1}\left(\frac{G}{Y}\right)$$

Or

$$\tan\phi\ \frac{B}{G} \qquad \phi-\tan^{-1}\left(\frac{B}{G}\right)$$

As the admittance, Y of a parallel RLC circuit is a complex quantity, the admittance corresponding to the general form of impedance Z = R + jX for series circuits will be written as Y = G - jB for parallel circuits where the real part G is the conductance and the imaginary part jB is the susceptance. In polar form this will be given as:

$$Y=G+jB=\sqrt{G^2+B^2}\ \angle\tan^{-1}\frac{B}{G}.$$

References

- Brief-introduction-to-circuits: electricaleasy.com, Retrieved 2 July, 2019
- Different-electronic-circuit-design-process: elprocus.com, Retrieved 14 April, 2019
- Difference-between-analog-circuit-and-digital-circuit: elprocus.com, Retrieved 18 June, 2019
- Different-electronic-circuit-design-process: elprocus.com, Retrieved 8 February, 2019
- Electronic-circuit-symbols: circuitstoday.com, Retrieved 26 June, 2019

- Basics-of-pcb: circuitdigest.com, Retrieved 17 August, 2019
- What-is-an-electric-circuit, Retrieved 7 January, 2019
- Dc-circuit: circuitglobe.com, Retrieved 13 March, 2019
- Ac-circuit-theory: circuitdigest.com, Retrieved 19 August, 2019
- ACcircuits: physics.bu.edu, Retrieved 9 May, 2019
- What-is-an-rc-resistor-capacitor-circuit: circuitglobe.com, Retrieved 25 January, 2019
- Series-and-parallel-lc-circuit-resonance: elprocus.com, Retrieved 5 March, 2019
- Rl-series-circuit: electrical4u.com, Retrieved 29 January, 2019
- Rl-parallel-circuit: electrical4u.com, Retrieved 2 May, 2019
- Series-circuit, accircuits: electronics-tutorials.ws, Retrieved 1 February, 2019
- Parallel-circuit, accircuits: electronics-tutorials.ws, Retrieved 17 April, 2019

2

Basic Elements of Circuits

There are various components which can be present in a circuit. These are divided into two categories, passive elements and active elements. Some of the passive elements are resistors, capacitors, inductors and diodes. This chapter discusses in detail these basic elements of circuits, along with active sources such as voltage sources and current sources.

Resistors

Resistors are electronic components which have a specific, never-changing electrical resistance. The resistor's resistance limits the flow of electrons through a circuit.

They are passive components, meaning they only consume power (and can't generate it). Resistors are usually added to circuits where they complement active components like op-amps, microcontrollers, and other integrated circuits. Commonly resistors are used to limit current, divide voltages, and pull-up I/O lines.

Resistor Units

The electrical resistance of a resistor is measured in ohms. The symbol for an ohm is the greek capital-omega: Ω. The (somewhat roundabout) definition of 1Ω is the resistance between two points where 1 volt (1V) of applied potential energy will push 1 ampere (1A) of current.

As SI units go, larger or smaller values of ohms can be matched with a prefix like kilo, mega or giga, to make large values easier to read. It's very common to see resistors in the kilohm (kΩ) and megaohm (MΩ) range (much less common to see miliohm (mΩ) resistors). For example, a 4,700Ω resistor is equivalent to a 4.7kΩ resistor, and a 5,600,000Ω resistor can be written as 5,600kΩ or (more commonly as) 5.6MΩ.

Schematic Symbol

All resistors have two terminals, one connection on each end of the resistor. When modelled on a schematic, a resistor will show up as one of these two symbols:

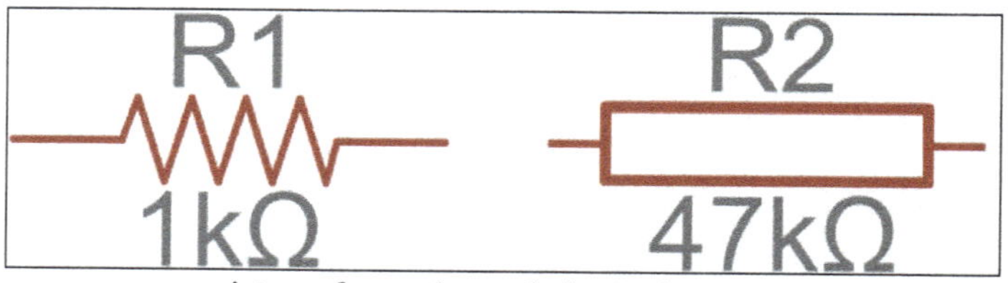

Two common resistor schematic symbols. R1 is an American-style 1kΩ resistor, and R2 is an international-style 47kΩ resistor.

The terminals of the resistor are each of the lines extending from the squiggle (or rectangle). Those are what connect to the rest of the circuit.

The resistor circuit symbols are usually enhanced with both a resistance value and a name. The value, displayed in ohms, is obviously critical for both evaluating and actually constructing the circuit. The name of the resistor is usually an *R* preceding a number. Each resistor in a circuit should have a unique name/number. For example, here's a few resistors in action on a 555 timer circuit:

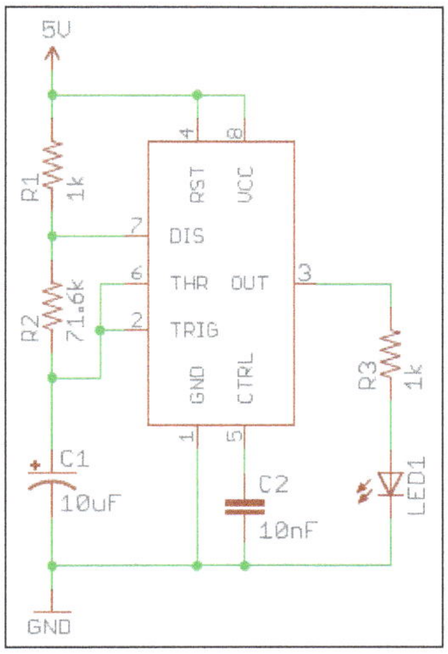

In this circuit, resistors play a key role in setting the frequency of the 555 timer's output. Another resistor (R3) limits the current through an LED.

Resistor Composition

Resistors can be constructed out of a variety of materials. Most common, modern resistors are made out of either a carbon, metal, or metal-oxide film. In these resistors, a thin film of conductive (though still resistive) material is wrapped in a helix around and covered by an insulating material. Most of the standard, no-frills, through-hole resistors will come in a carbon-film or metal-film composition.

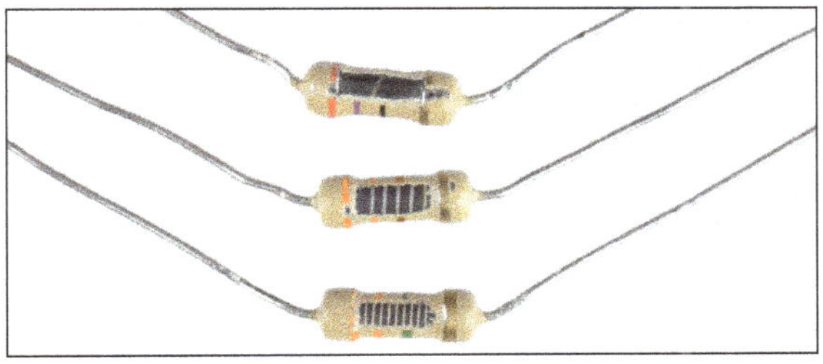

Peek inside the guts of a few carbon-film resistors. Resistance values from top to bottom

Other through-hole resistors might be wire wound or made of super-thin metallic foil. These resistors are usually more expensive, higher-end components specifically chosen for their unique characteristics like a higher power-rating, or maximum temperature range.

Surface-mount resistors are usually either thick or thin-film variety. Thick-film is usually cheaper but less precise than thin. In both resistor types, a small film of resistive metal alloy is sandwiched between a ceramic base and glass/epoxy coating, and then connected to the terminating conductive edges.

Special Resistor Packages

There are a variety of other, special-purpose resistors out there. Resistors may come in pre-wired packs of five-or-so resistor arrays. Resistors in these arrays may share a common pin, or be set up as voltage dividers.

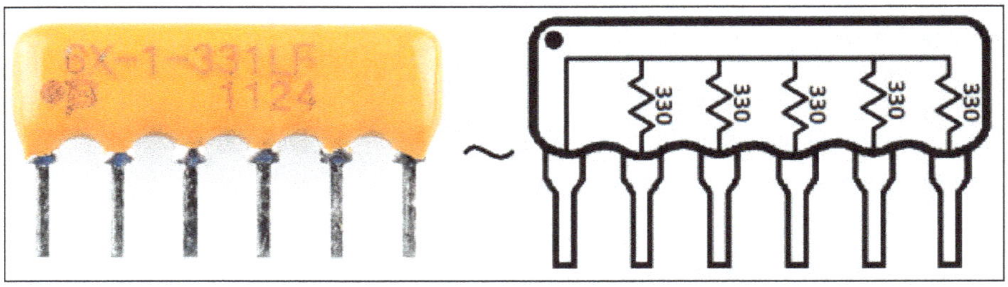

An array of five 330Ω resistors, all tied together at one end.

Variable Resistors

Resistors don't have to be static either. Variable resistors, known as rheostats, are resistors which can be adjusted between a specific range of values. Similar to the rheostat is the potentiometer. Pots connect two resistors internally, in series, and adjust a center tap between them creating an adjustable voltage divider. These variable resistors are often used for inputs, like volume knobs, which need to be adjustable.

A smattering of potentiometers. From top-left, clockwise: a standard 10k trimpot, 2-axis joystick, softpot, slide pot, classic right-angle, and a breadboard friendly 10k trimpot.

Decoding Resistor Markings

Though they may not display their value outright, most resistors are marked to show what their resistance is. PTH resistors use a color-coding system (which really adds some flair to circuits), and SMD resistors have their own value-marking system.

Decoding the Color Bands

Through-hole, axial resistors usually use the color-band system to display their value. Most of these resistors will have four bands of color circling the resistor, though you will also find five band and six band resistors.

Four Band Resistors

In the standard four band resistors, the first two bands indicate the two most-significant digits of the resistor's value. The third band is a weight value, which multiplies the two significant digits by a power of ten.

The final band indicates the tolerance of the resistor. The tolerance explains how much more or less the *actual* resistance of the resistor can be compared to what its nominal value is. No resistor is made to perfection, and different manufacturing processes will result in better or worse tolerances. For example, a 1kΩ resistor with 5% tolerance could actually be anywhere between 0.95kΩ and 1.05kΩ.

The last, tolerance band is often clearly separated from the value bands, and usually it'll either be silver or gold.

Five and Six Band Resistors

Five band resistors have a third significant digit band between the first two bands and the multiplier band. Five band resistors also have a wider range of tolerances available.

Six band resistors are basically five band resistors with an additional band at the end that indicates the temperature coefficient. This indicates the expected change in resistor value as the temperature changes in degrees Celsius. Generally these temperature coefficient values are extremely small, in the ppm range.

Decoding Resistor Color Bands

When decoding the resistor color bands, consult a resistor color code table like the one below. For the first two bands, find that color's corresponding digit value. The 4.7kΩ resistor shown here has color bands of yellow and violet to begin - which have digit values of 4 and 7 (47). The third band of the 4.7kΩ is red, which indicates that the 47 should be multiplied by 10^2 (or 100). 47 times 100 is 4,700.

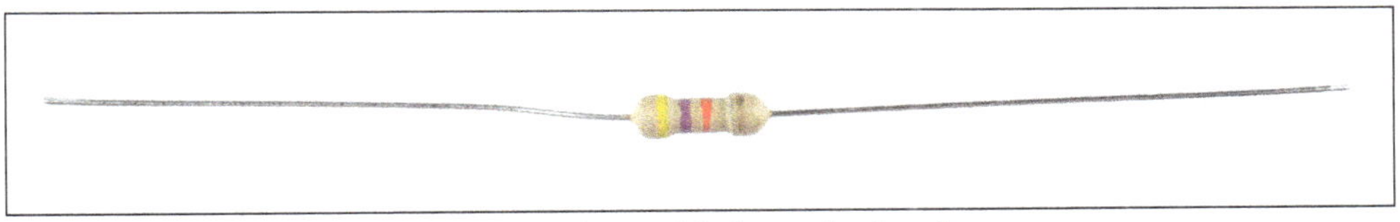

4.7kΩ resistor with four color bands

There are a handful of(sometimes unsavory) mnemonics out there to help remember the resistor color code. A good one, which spells out the difference between *b*lack and *b*rown is:

"Big brown rabbits often yield great big vocal groans when gingerly snapped."

Resistor Color Code Table

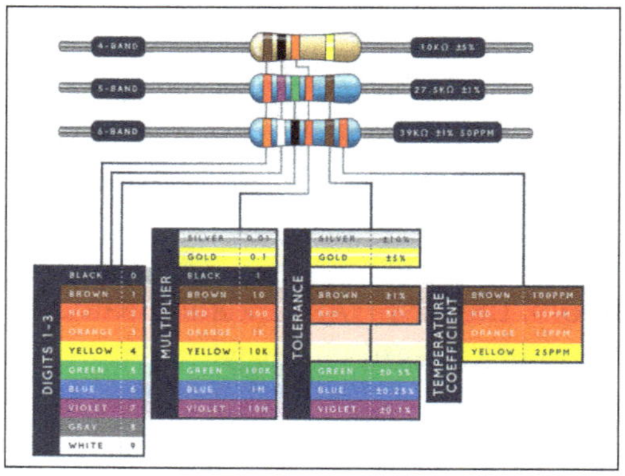

Decoding Surface-mount Markings

SMD resistors, like those in 0603 or 0805 packages, have their own way of displaying their value. There are a few common marking methods you'll see on these resistors. They'll usually have three to four characters - numbers or letters - printed on top of the case.

If the three characters you're seeing are *all numbers*, you're probably looking at an E24 marked resistor. These markings actually share some similarity with the color-band system used on the PTH resistors. The first two numbers represent the first two most-significant digits of the value the last number represents a magnitude.

In the above example picture, resistors are marked *104*, *105*, *205*, *751*, and *754*. The resistor marked with *104* should be 100kΩ ($10x10^4$), *105* would be 1MΩ ($10x10^5$), and *205* is 2MΩ ($20x10^5$). *751* is 750Ω ($75x10^1$), and *754* is 750kΩ ($75x10^4$).

Another common coding system is E96, and it's the most cryptic of the bunch. E96 resistors will be marked with three characters -- two numbers at the beginning and a letter at the end. The two numbers tell you the first *three* digits of the value, by corresponding to one of the not-so-obvious values on this lookup table.

Code	Value	Code	Value	Code	Value	Code	Value	Code	Value	Code	Value
01	100	17	147	33	215	49	316	65	464	81	681
02	102	18	150	34	221	50	324	66	475	82	698
03	105	19	154	35	226	51	332	67	487	83	715

04	107	20	158	36	232	52	340	68	499	84	732
05	110	21	162	37	237	53	348	69	511	85	750
06	113	22	165	38	243	54	357	70	523	86	768
07	115	23	169	39	249	55	365	71	536	87	787
08	118	24	174	40	255	56	374	72	549	88	806
09	121	25	178	41	261	57	383	73	562	89	825
10	124	26	182	42	267	58	392	74	576	90	845
11	127	27	187	43	274	59	402	75	590	91	866
12	130	28	191	44	280	60	412	76	604	92	887
13	133	29	196	45	287	61	422	77	619	93	909
14	137	30	200	46	294	62	432	78	634	94	931
15	140	31	205	47	301	63	442	79	649	95	953
16	143	32	210	48	309	64	453	80	665	96	976

The letter at the end represents a multiplier, matching up to something on this table:

Letter	Multiplier	Letter	Multiplier	Letter	Multiplier
Z	0.001	A	1	D	1000
Y or R	0.01	B or H	10	E	10000
X or S	0.1	C	100	F	100000

So a *01C* resistor is our good friend, 10kΩ (100x100), *01B* is 1kΩ (100x10), and *01D* is 100kΩ. Those are easy, other codes may not be. *85A* from the picture above is 750Ω (750x1) and *30C* is actually 20kΩ.

Power Rating

The power rating of a resistor is one of the more hidden values.

Power is the rate at which energy is transformed into something else. It's calculated by multiplying the voltage difference across two points by the current running between them, and is measured in units of a watt (W). Light bulbs, for example, power electricity into light. But a resistor can only turn electrical energy running through it into heat. Heat isn't usually a nice playmate with electronics; too much heat leads to smoke, sparks, and fire.

Every resistor has a specific maximum power rating. In order to keep the resistor from heating up too much, it's important to make sure the power across a resistor is kept under it's maximum rating. The power rating of a resistor is measured in watts, and it's usually somewhere between 1/8W (0.125W) and 1W. Resistors with power ratings of more than 1W are usually referred to as power resistors, and are used specifically for their power dissipating abilities.

Finding a Resistor's Power Rating

A resistor's power rating can usually be deduced by observing its package size. Standard through-hole resistors usually come with ¼W or ½W ratings. More special purpose, power resistors might actually list their power rating on the resistor.

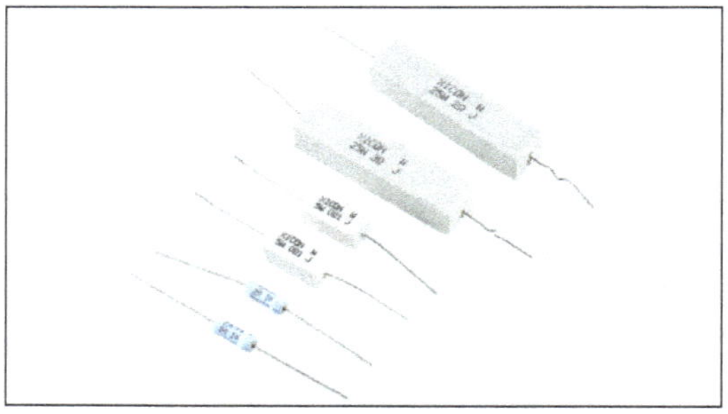

These power resistors can handle a lot more power before they blow. From top-right to bottom-left there are examples of 25W, 5W and 3W resistors, with values of 2Ω, 3Ω 0.1Ω and 22kΩ. Smaller power-resistors are often used to sense current.

The power ratings of surface mount resistors can usually be judged by their size as well. Both 0402 and 0603-size resistors are usually rated for 1/16W, and 0805's can take 1/10W.

Measuring Power across a Resistor

Power is usually calculated by multiplying voltage and current (P = IV). But, by applying Ohm's law, we can also use the resistance value in calculating power. If we know the current running through a resistor, we can calculate the power as:

$$p = 1^2 . R$$

Or, if we know the voltage across a resistor, the power can be calculated as:

$$P = \frac{V^2}{R}$$

Series and Parallel Resistors

Resistors are paired together all the time in electronics, usually in either a series or parallel circuit. When resistors are combined in series or parallel, they create a total resistance, which can be calculated using one of two equations. Knowing how resistor values combine comes in handy if you need to create a specific resistor value.

Series Resistors

When connected in series resistor values simply add up.

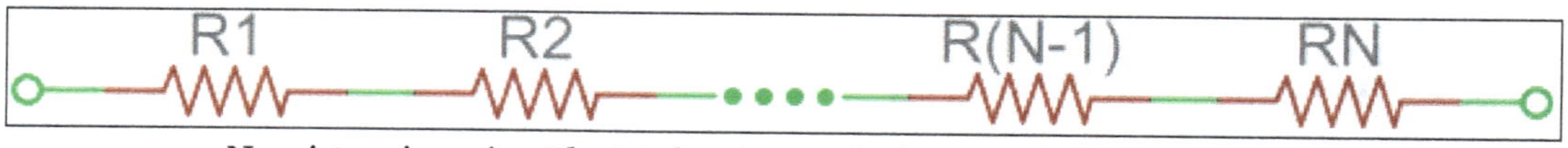

N resistors in series. The total resistance is the sum of all series resistors.

$$R_{tot} = R_1 + R_2 \ldots + R_{N-1} + R_N$$

So, for example, if you just *have to have* a 12.33kΩ resistor, seek out some of the more common resistor values of 12kΩ and 330Ω, and butt them up together in series.

Parallel Resistors

Finding the resistance of resistors in parallel isn't quite so easy. The total resistance of *N* resistors in parallel is the inverse of the sum of all inverse resistances.

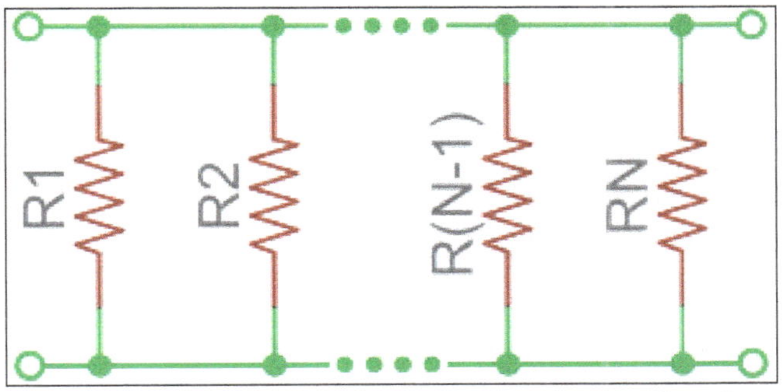

N resistors in parallel. To find the total resistance, invert each resistance value, add them up, and then invert that.

$$\frac{1}{R_{tot}} = \frac{1}{R_1} + \frac{1}{R_2} \ldots + \frac{1}{R_{N-1}} + \frac{1}{R_N}$$

(The inverse of resistance is actually called conductance, so put more succinctly: the *conductance* of parallel resistors is the sum of each of their conductance's).

As a special case of this equation: if you have just two resistors in parallel, their total resistance can be calculated with this slightly-less-inverted equation:

$$R_{tot} = \frac{R_1 . R_2}{R_1 + R_2}$$

As an even *more special* case of that equation, if you have two parallel resistors of equal value the total resistance is half of their value. For example, if two 10kΩ resistors are in parallel, their total resistance is 5kΩ.

A shorthand way of saying two resistors are in parallel is by using the parallel operator: ||. For example, if R_1 is in parallel with R_2, the conceptual equation could be written as $R_1||R_2$. Much cleaner, and hides all those nasty fractions.

Resistor Networks

As a special introduction to calculating total resistances, electronics teachers just *love* to subject their students to finding that of crazy, convoluted resistor networks.

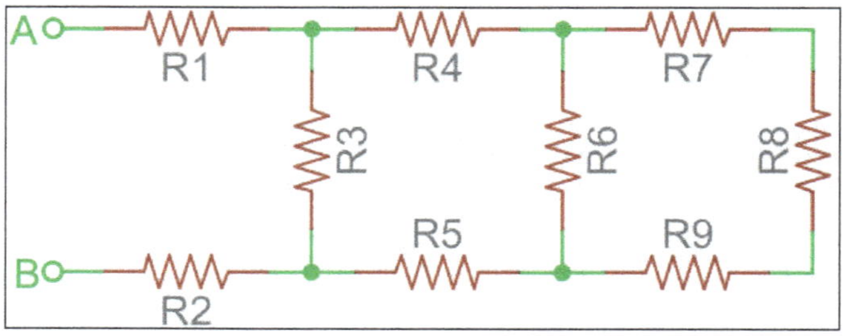

To solve such a problem, start at the back-end of the circuit and simplify towards the two terminals. In this case R_7, $R_8$and R_9 are all in series and can be added together. Those three resistors are in parallel with R_6, so those four resistors could be turned into one with a resistance of $R_6||(R_7+R_8+R_9)$. Making our circuit:

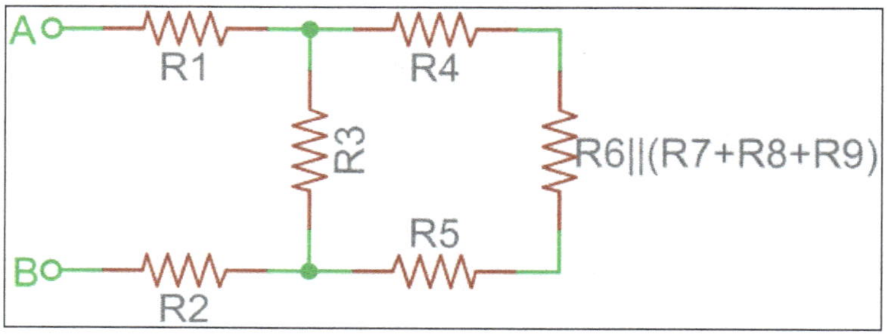

Now the four right-most resistors can be simplified even further. R_4, R_5 and our conglomeration of R_6 - R_9 are all in series and can be added. Then those series resistors are all in parallel with R_3.

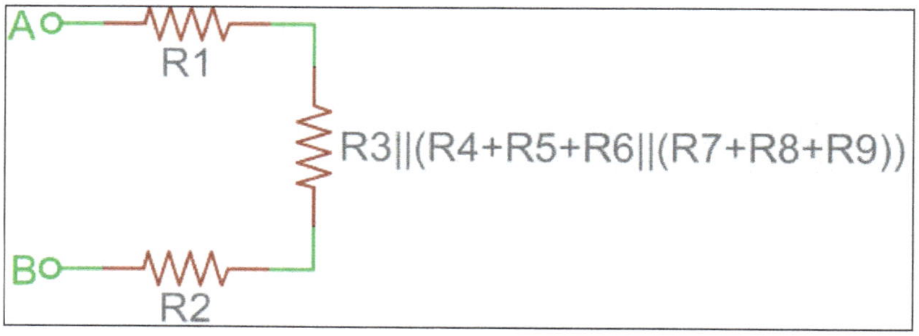

And that's just three series resistors between the *A* and *B* terminals. Add 'em on up! So the total resistance of that circuit is: $R_1+R_2+R_3||(R_4+R_5+R_6||(R_7+R_8+R_9))$.

Resistors exist in just about every electronic circuit ever. Here are a few examples of circuits, which heavily depend on our resistor friends.

LED Current Limiting

Resistors are key in making sure LEDs don't blow up when power is applied. By connecting a resistor in series with an LED, current flowing through the two components can be limited to a safe value.

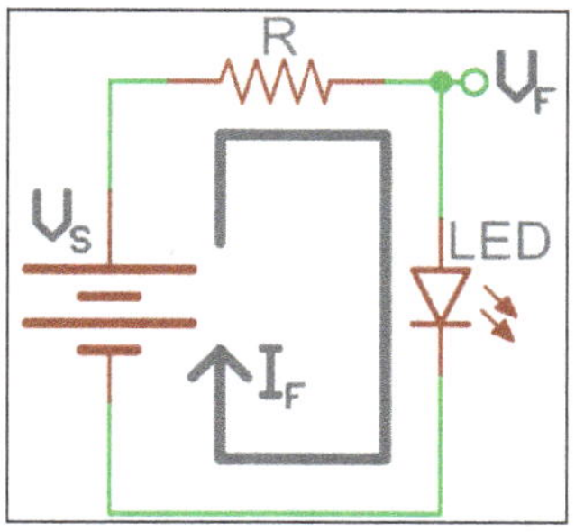

When sizing out a current-limiting resistor, look for two characteristic values of the LED: the typical forward voltage, and the maximum forward current. The typical forward voltage is the voltage which is required to make an LED light up, and it varies (usually somewhere between 1.7V and 3.4V) depending upon the color of the LED. The maximum forward current is usually around 20mA for basic LEDs; continuous current through the LED should always be equal to or less than that current rating.

Once you've gotten a hold of those two values, you can size up a current-limiting resistor with this equation:

$$R = \frac{V_S - V_F}{I_F}$$

V_S is the source voltage usually a battery or power supply voltage. V_F and I_F are the LED's forward voltage and the desired current that runs through it.

For example, assume you have a 9V battery to power an LED. If your LED is red, it might have a forward voltage around 1.8V. If you want to limit the current to 10mA, use a series resistor of about 720Ω.

$$R = \frac{V_S - V_F}{I_F} = \frac{9 - 1.8}{0.010} = 720\Omega$$

Voltage Dividers

A voltage divider is a resistor circuit which turns a large voltage into a smaller one. Using just two resistors in series, an output voltage can be created that's a fraction of the input voltage.

Here's the voltage divider circuit:

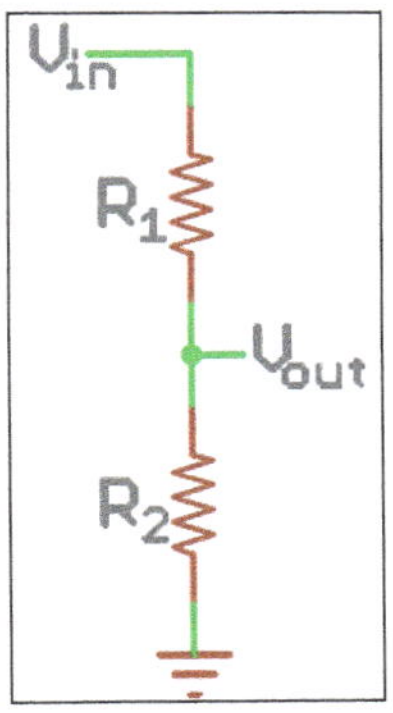

Two resistors, R_1 and R_2, are connected in series and a voltage source (V_{in}) is connected across them. The voltage from V_{out} to GND can be calculated as:

$$V_{out} = V_{in} \cdot \frac{R_2}{R_1 + R_2}$$

For example, if R_1 was 1.7kΩ and R_2 was 3.3kΩ, a 5V input voltage could be turned into 3.3V at the V_{out} terminal.

Voltage dividers are very handy for reading resistive sensors, like photocells, flex sensors, and force-sensitive resistors. One half of the voltage divider is the sensor, and the part is a static resistor. The output voltage between the two components is connected to an analog-to-digital converter on a microcontroller (MCU) to read the sensor's value.

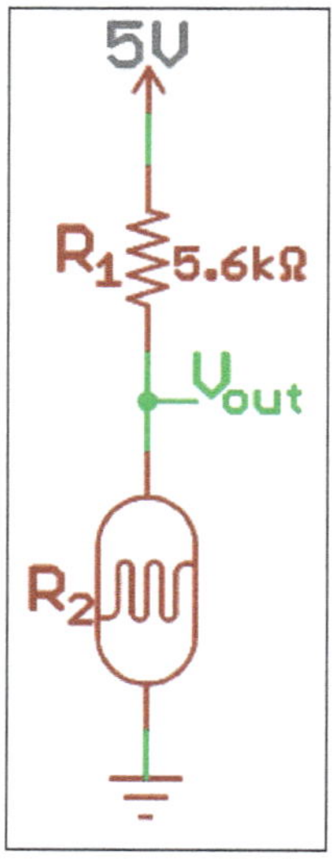

Here a resistor R_1 and a photocell create a voltage divider to create a variable voltage output.

Pull-up Resistors

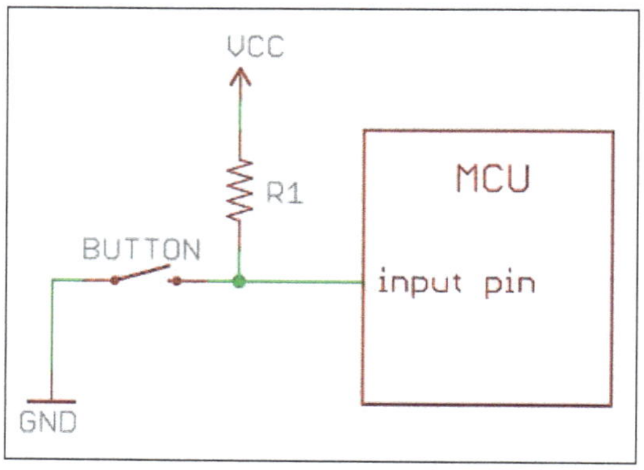

A pull-up resistor is used when you need to bias a microcontroller's input pin to a known state. One end of the resistor is connected to the MCU's pin, and the other end is connected to a high voltage (usually 5V or 3.3V).

Without a pull-up resistor, inputs on the MCU could be left *floating*. There's no guarantee that a floating pin is either high (5V) or low (0V).

Pull-up resistors are often used when interfacing with a button or switch input. The pull-up resistor can bias the input-pin when the switch is open. And it will protect the circuit from a short when the switch is closed.

In the circuit above, when the switch is open the MCU's input pin is connected through the resistor to 5V. When the switch closes, the input pin is connected directly to GND.

Resistance and Temperature

The resistance of a resistor isn't constant, even if it's a certain material of a fixed length and area: it steadily *increases* as the temperature increases. Why? The hotter a material, the more its atoms or ions jiggle about and the harder it is for electrons to wriggle through, which translates into higher electrical resistance. Broadly speaking, the resistivity of most materials increases linearly with temperature (so if you increase the temperature by 10 degrees, the resistivity increases by a certain amount, and if you increase it by another 10 degrees, the resistivity rises by the same amount again). If you *cool* a material, you lower its resistivity—and if you cool it to an extremely low temperature, you can sometimes make the resistivity disappear altogether, in a phenomenon known as superconductivity.

Types of Resistor

There are many different Types of Resistor available for the electronics constructor to choose from, from very small surface mount chip resistors up to large wire wound power resistors.

Carbon Resistors are the most common type of Composition Resistors. Carbon resistors are a cheap general purpose resistor used in electrical and electronic circuits. Their resistive element is manufactured from a mixture of finely ground carbon dust or graphite (similar to pencil lead) and a non-conducting ceramic (clay) powder to bind it all together.

Carbon Resistor

The ratio of carbon dust to ceramic (conductor to insulator) determines the overall resistive value of the mixture and the higher the ratio of carbon, the lower the overall resistance. The mixture is moulded into a cylindrical shape with metal wires or leads are attached to each end to provide the electrical connection as shown, before being coated with an outer insulating material and colour coded markings to denote its resistive value.

Carbon Resistor

The Carbon Composite Resistor is a low to medium type power resistor which has a low inductance making them ideal for high frequency applications but they can also suffer from noise and stability when hot. Carbon composite resistors are generally prefixed with a "CR" notation (eg, CR10kΩ) and are available in E6 (± 20% tolerance (accuracy)), E12 (± 10% tolerance) and E24 (± 5% tolerance) packages with power ratings from 0.250 or 1/4 of a Watt up to 5 Watts.

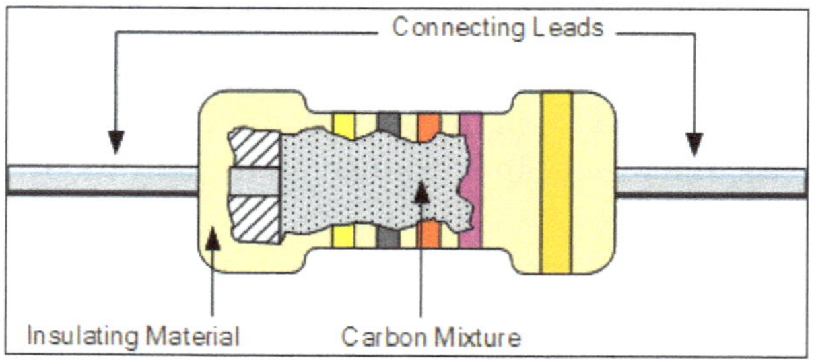

Carbon composite resistor types are very cheap to make and are therefore commonly used in electrical circuits. However, due to their manufacturing process carbon type resistors have very large tolerances so for more precision and high value resistances, film type resistors are used instead.

Film Type Resistors

The generic term "Film Resistor" consist of *Metal Film*, *Carbon Film* and *Metal Oxide Film* resistor types, which are generally made by depositing pure metals, such as nickel, or an oxide film, such as tin-oxide, onto an insulating ceramic rod or substrate.

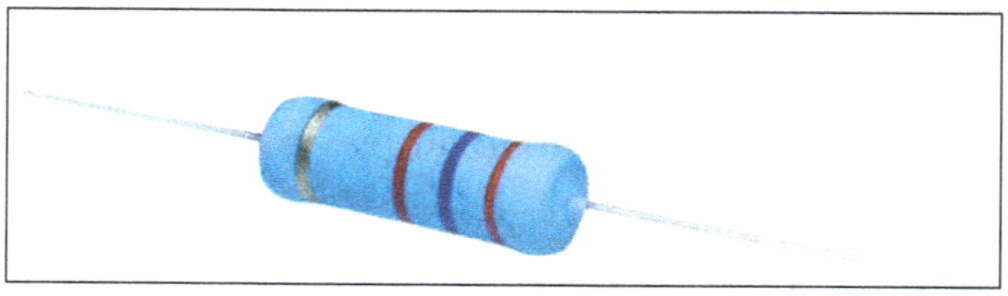

Film Resistor

The resistive value of the resistor is controlled by increasing the desired thickness of the deposited film giving them the names of either "thick-film resistors" or "thin-film resistors".

Once deposited, a laser is used to cut a high precision spiral helix groove type pattern into this film. The cutting of the film has the effect of increasing the conductive or resistive path, a bit like taking a long length of straight wire and forming it into a coil.

This method of manufacture allows for much closer tolerance resistors (1% or less) as compared to the simpler carbon composition types. The tolerance of a resistor is the difference between the preferred value (i.e. 100 ohms) and its actual manufactured value i.e. 103.6 ohms, and is expressed as a percentage, for example 5%, 10% etc. and in our example the actual tolerance is 3.6%. Film type resistors also achieve a much higher maximum ohmic value compared to other types and values in excess of 10MΩ (10 Million Ohms) are available.

Film Resistor

Metal Film Resistors have much better temperature stability than their carbon equivalents, lower noise and are generally better for high frequency or radio frequency applications. Metal Oxide

Resistors have better high surge current capability with a much higher temperature rating than the equivalent metal film resistors.

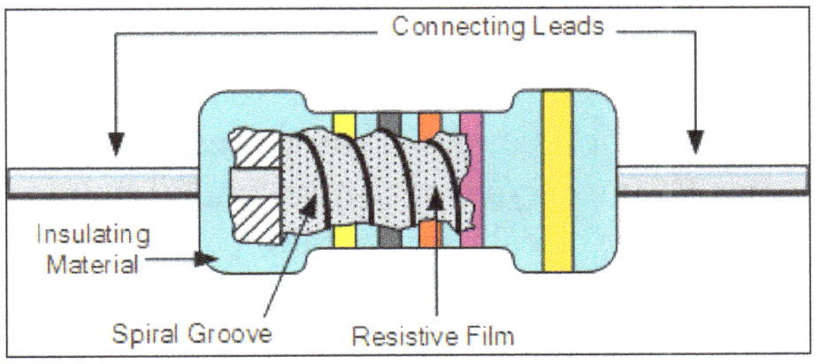

Another type of film resistor commonly known as a Thick Film Resistor is manufactured by depositing a much thicker conductive paste of ceramic and metal, called Cermet, onto an alumina ceramic substrate. Cermet resistors have similar properties to metal film resistors and are generally used for making small surface mount chip type resistors, multi-resistor networks in one package for pcb's and high frequency resistors. They have good temperature stability, low noise, and good voltage ratings but low surge current properties.

Metal Film Resistors are prefixed with a "MFR" notation (e.g. MFR100kΩ) and a CF for Carbon Film types. Metal film resistors are available in E24 (±5% & ±2% tolerances), E96 (±1% tolerance) and E192(±0.5%, ±0.25% & ±0.1% tolerances) packages with power ratings of 0.05 (1/20th) of a Watt up to 1/2 Watt. Generally speaking Film resistors and especially metal film resistors are precision low power components.

Wire Wound Type of Resistor

Another type of resistor, called a Wire wound Resistor, is made by winding a thin metal alloy wire (Nichrome) or similar wire onto an insulating ceramic former in the form of a spiral helix similar to the film resistor above.

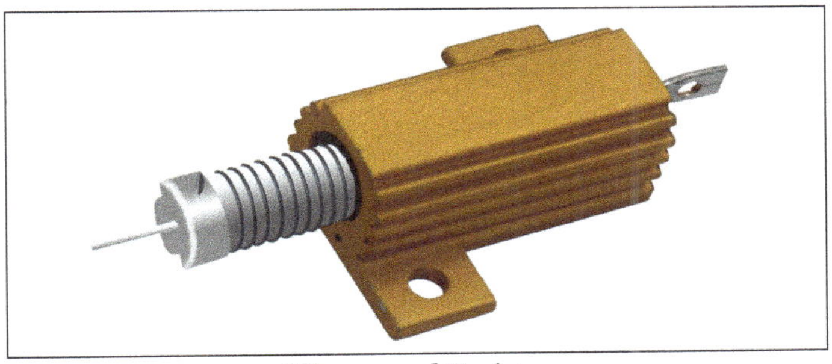

Wire wound Resistor

These types of resistor are generally only available in very low ohmic high precision values (from 0.01Ω to 100kΩ) due to the gauge of the wire and number of turns possible on the former making them ideal for use in measuring circuits and Wheatstone bridge type applications.

They are also able to handle much higher electrical currents than other resistors of the same ohmic value with power ratings in excess of 300 Watts. These high power resistors are moulded or pressed into an aluminium heat sink body with fins attached to increase their overall surface area to promote heat loss and cooling.

These special types of resistor are called "Chassis Mounted Resistors" because they are designed to be physically mounted onto heat sinks or metal plates to further dissipate the generated heat. The mounting of the resistor onto a heat sink increases their current carrying capabilities even further.

Another type of wire wound resistor is the Power Wire wound Resistor. These are high temperature, high power non-inductive resistor types generally coated with a vitreous or glass epoxy enamel for use in resistance banks or DC motor/servo control and dynamic braking applications. They can even be used as low wattage space or cabinet heaters.

The non-inductive resistance wire is wound around a ceramic or porcelain tube covered with mica to prevent the alloy wires from moving when hot. Wire wound resistors are available in a variety of resistance and power ratings with one main use of power wire wound resistor is in the electrical heating elements of an electric fire which converts the electrical current flowing through it into heat with each element dissipating up to 1000 Watts, (1kW) of energy.

Because the wire of standard wire wound resistors is wound into a coil inside the resistors body, it acts like an inductor causing them to have inductance as well as resistance. This affects the way the resistor behaves in AC circuits by producing a phase shift at high frequencies especially in the larger size resistors. The length of the actual resistance path in the resistor and the leads contributes inductance in series with the "apparent" DC resistance resulting in an overall impedance path of Z Ohms.

Impedance (Z) is the combined effect of resistance (R) and inductance (X), measured in ohms and for a series AC circuit is given as, $Z^2 = R^2 + X^2$.

When used in AC circuits this inductance value changes with frequency (inductive reactance, $X_L = 2\pi f L$) and therefore, the overall value of the resistor changes. Inductive reactance increases with frequency but is zero at DC (zero frequency). Then, wire wound resistors must not be designed or used in AC or amplifier type circuits where the frequency across the resistor changes. However, special non-inductive wire wound resistors are also available.

Wire Wound Resistor

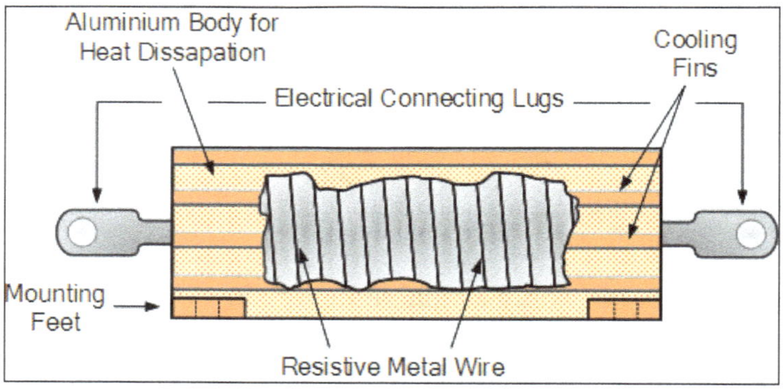

Wire wound resistor types are prefixed with a "WH" or "W" notation (e.g. WH10Ω) and are available in the WH aluminium clad package (±1%, ±2%, ±5% and ±10% tolerance) or the W vitreous enamelled package (±1%, ±2% and ±5% tolerance) with power ratings from 1W to 300W or more.

Resistor Applications

There is a huge variation in fields of applications for resistors; from precision components in digital electronics, till measurement devices for physical quantities.

Resistors in Series and Parallel

In electronic circuits, resistors are very often connected in series or in parallel. A circuit designer might for example combine several resistors with standard values (E-series) to reach a specific resistance value. For series connection, the current through each resistor is the same and the equivalent resistance is equal to the sum of the individual resistors. For parallel connection, the voltage through each resistor is the same, and the inverse of the equivalent resistance is equal to the sum of the inverse values for all parallel resistors.

Measure Electrical Current (Shunt Resistor)

Electrical current can be calculated by measuring the voltage drop over a precision resistor with a known resistance, which is connected in series with the circuit. The current is calculated by using Ohm's law. This is a called an ammeter or shunt resistor. Usually this is a high precision manganin resistor with a low resistance value.

Resistors for LEDs

LED lights need a specific current to operate. A too low current will not light up the LED, while a too high current might burn out the device. Therefore, they are often connected in series with resistors. These are called ballast resistors and passively regulate the current in the circuit.

Blower Motor Resistor

In cars the air ventilation system is actuated by a fan that is driven by the blower motor. A special resistor is used to control the fan speed. This is called the blower motor resistor. Different designs are in use. One design is a series of different size wire wound resistors for each fan speed. Another design incorporates a fully integrated circuit on a printed circuit board.

Capacitor

A capacitor (originally known as a condenser) is a passive two-terminal electrical component used to store energy electrostatically in an electric field. The forms of practical capacitors vary widely, but all contain at least two electrical conductors (plates) separated by a dielectric (i.e., insulator). The conductors can be thin films of metal, aluminum foil or disks, etc. The 'nonconducting' dielectric acts to increase the capacitor's charge capacity. A dielectric can be glass, ceramic, plastic film, air, paper, mica, etc. Capacitors are widely used as parts of electrical circuits in many common electrical devices. Unlike a resistor, a capacitor does not dissipate energy. Instead, a capacitor stores energy in the form of an electrostatic field between its plates.

When there is a potential difference across the conductors (e.g., when a capacitor is attached across a battery), an electric field develops across the dielectric, causing positive charge (+Q) to collect on one plate and negative charge (-Q) to collect on the other plate. If a battery has been attached to a capacitor for a sufficient amount of time, no current can flow through the capacitor. However, if an accelerating or alternating voltage is applied across the leads of the capacitor, a displacement current can flow.

An ideal capacitor is characterized by a single constant value for its capacitance. Capacitance is expressed as the ratio of the electric charge (Q) on each conductor to the potential difference (V) between them. The SI unit of capacitance is the farad (F), which is equal to one coulomb per volt (1 C/V). Typical capacitance values range from about 1 pF (10^{-12} F) to about 1 mF (10^{-3} F).

The capacitance is greater when there is a narrower separation between conductors and when the conductors have a larger surface area. In practice, the dielectric between the plates passes a small amount of leakage current and also has an electric field strength limit, known as the breakdown voltage. The conductors and leads introduce an undesired inductance and resistance.

Capacitors are widely used in electronic circuits for blocking direct current while allowing alternating current to pass. In analog filter networks, they smooth the output of power supplies. In resonant circuits they tune radios to particular frequencies. In electric power transmission systems they stabilize voltage and power flow.

Symbols and Units

Circuit Symbols

There are two common ways to draw a capacitor in a schematic. They always have two terminals, which go on to connect to the rest of the circuit. The capacitors symbol consists of two parallel lines, which are either flat or curved; both lines should be parallel to each other, close, but not touching (this is actually representative of how the capacitor is made.

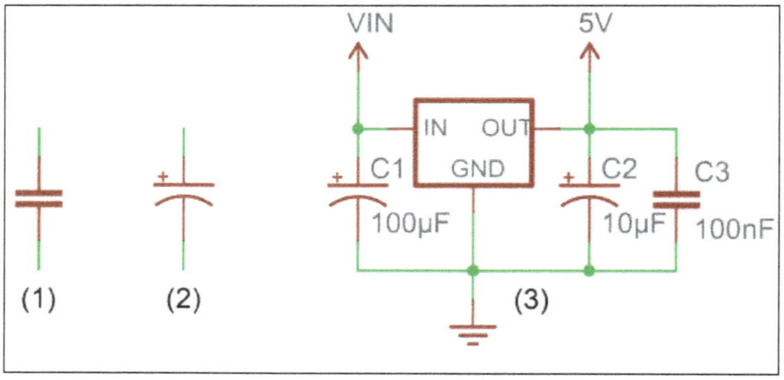

(1) and (2) are standard capacitor circuit symbols. (3) is an example of capacitors symbols in action in a voltage regulator circuit.

The symbol with the curved line (#2 in the photo above) indicates that the capacitor is polarized, meaning it's probably an electrolytic capacitor.

Each capacitor should be accompanied by a name - C1, C2, etc. and a value. The value should indicate the capacitance of the capacitor; how many farads it has.

Capacitance Units

Not all capacitors are created equal. Each capacitor is built to have a specific amount of capacitance. The capacitance of a capacitor tells you how much charge it can store, more capacitance means more capacity to store charge. The standard unit of capacitance is called the farad, which is abbreviated *F*.

It turns out that a farad is a *lot* of capacitance, even 0.001F (1 milifarad - 1mF) is a big capacitor. Usually you'll see capacitors rated in the pico- (10^{-12}) to microfarad (10^{-6}) range.

Prefix Name	Abbreviation	Weight	Equivalent Farads
Picofarad	pF	10^{-12}	0.000000000001 F
Nanofarad	nF	10^{-9}	0.000000001 F
Microfarad	µF	10^{-6}	0.000001 F
Milifarad	mF	10^{-3}	0.001 F
Kilofarad	kF	10^{3}	1000 F

When you get into the farad to kilofarad range of capacitance, you start talking about special caps called *super* or *ultra*-capacitors.

Capacitor Theory

The schematic symbol for a capacitor actually closely resembles how it's made. A capacitor is created out of two metal plates and an insulating material called a dielectric. The metal plates are placed very close to each other, in parallel, but the dielectric sits between them to make sure they don't touch.

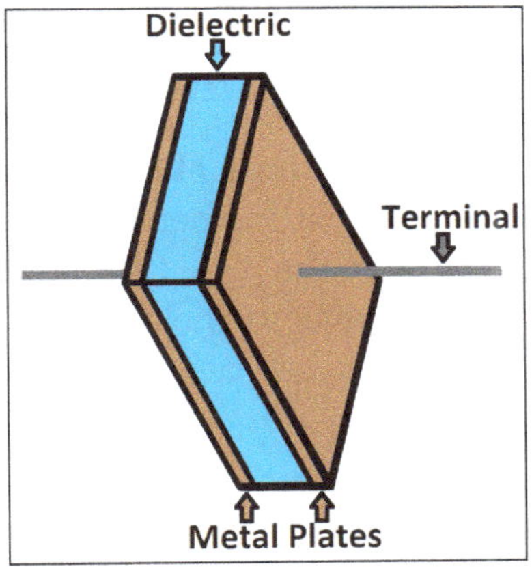

Standard capacitor sandwich: two metal plates separated by an insulating dielectric.

The dielectric can be made out of all sorts of insulating materials: paper, glass, rubber, ceramic, plastic, or anything that will impede the flow of current.

The plates are made of a conductive material: aluminum, tantalum, silver, or other metals. They're each connected to a terminal wire, which is what eventually connects to the rest of the circuit.

The capacitance of a capacitor - how many farads it has - depends on how it's constructed. More capacitance requires a larger capacitor. Plates with more overlapping surface area provide more capacitance, while more distance between the plates means less capacitance. The material of the dielectric even has an effect on how many farads a cap has. The total capacitance of a capacitor can be calculated with the equation:

$$C = \varepsilon r \frac{A}{4\pi d}$$

Where ε_r is the dielectric's relative permittivity (a constant value determined by the dielectric material), A is the amount of area the plates overlap each other, and d is the distance between the plates.

Working of Capacitor

Electric current is the flow of electric charge, which is what electrical components harness to light up, or spin, or do whatever they do. When current flows into a capacitor, the charges get "stuck" on the plates because they can't get past the insulating dielectric. Electrons -- negatively charged particles -- are sucked into one of the plates, and it becomes overall negatively charged. The large mass of negative charges on one plate pushes away like charges on the other plate, making it positively charged.

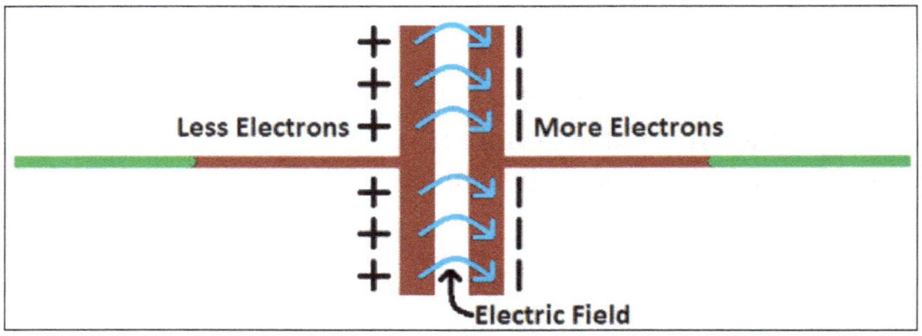

The positive and negative charges on each of these plates attract each other, because that's what opposite charges do. But, with the dielectric sitting between them, as much as they want to come together, the charges will forever be stuck on the plate (until they have somewhere else to go). The stationary charges on these plates create an electric field, which influence electric potential energy and voltage. When charges group together on a capacitor like this, the cap is storing electric energy just as a battery might store chemical energy.

Charging and Discharging

When positive and negative charges coalesce on the capacitor plates, the capacitor becomes charged. A capacitor can retain its electric field -- hold its charge -- because the positive and negative charges on each of the plates attract each other but never reach each other.

At some point the capacitor plates will be so full of charges that they just can't accept any more.

There are enough negative charges on one plate that they can repel any others that try to join. This is where the capacitance (farads) of a capacitor comes into play, which tells you the maximum amount of charge the cap can store.

If a path in the circuit is created, which allows the charges to find another path to each other, they'll leave the capacitor, and it will discharge.

For example, in the circuit below, a battery can be used to induce an electric potential across the capacitor. This will cause equal but opposite charges to build up on each of the plates, until they're so full they repel any more current from flowing. An LED placed in series with the cap could provide a path for the current, and the energy stored in the capacitor could be used to briefly illuminate the LED.

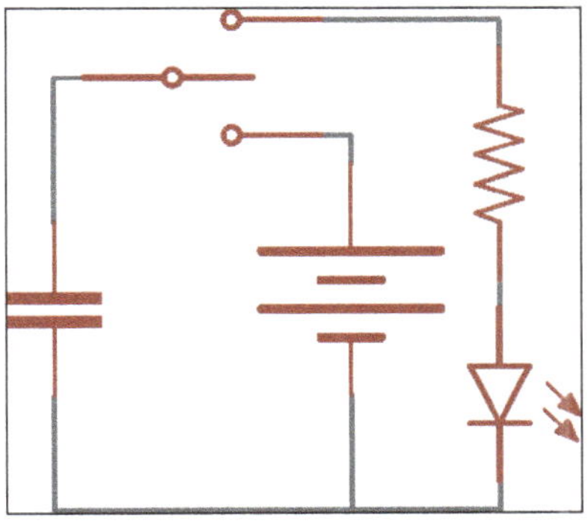

Calculating Charge, Voltage and Current

A capacitor's capacitance - how many farads it has - tells you how much charge it can store. How much charge a capacitor is *currently* storing depends on the potential difference (voltage) between its plates. This relationship between charge, capacitance, and voltage can be modelled with this equation:

$$Q = CV$$

Charge (Q) stored in a capacitor is the product of its capacitance (C) and the voltage (V) applied to it.

The capacitance of a capacitor should always be a constant, known value. So we can adjust voltage to increase or decrease the cap's charge. More voltage means more charge, less voltage less charge.

That equation also gives us a good way to define the value of one farad. One farad (F) is the capacity to store one unit of energy (coulombs) per every one volt.

Calculating Current

We can take the charge/voltage/capacitance equation a step further to find out how capacitance and voltage affect current, because current is the rate of flow of charge. The gist of a capacitor's relationship to voltage and current is this: the amount of current through a capacitor depends on both the capacitance and how quickly the voltage is rising or falling. If the voltage across a capacitor swiftly rises, a large positive current will be induced through the capacitor. A slower rise in voltage across a capacitor equates to a smaller current through it. If the voltage across a capacitor is steady and unchanging, no current will go through it.

The equation for calculating current through a capacitor is:

$$i = C\frac{dV}{dt}$$

The *dV/dt* part of that equation is a derivative (a fancy way of saying *instantaneous rate*) of voltage over time, it's equivalent to saying "how fast is voltage going up or down at this very moment". The big takeaway from this equation is that if voltage is steady, the derivative is zero, which means current is also zero. This is why current cannot flow through a capacitor holding a steady, DC voltage.

Types of Capacitors

There are all sorts of capacitor types out there, each with certain features and drawbacks which make it better for some applications than others.

When deciding on capacitor types there are a handful of factors to consider:

- Size - Size both in terms of physical volume and capacitance. It's not uncommon for a capacitor to be the largest component in a circuit. They can also be very tiny. More capacitance typically requires a larger capacitor.
- Maximum voltage - Each capacitor is rated for a maximum voltage that can be dropped across it. Some capacitors might be rated for 1.5V, others might be rated for 100V. Exceeding the maximum voltage will usually result in destroying the capacitor.
- Leakage current - Capacitors aren't perfect. Every cap is prone to leaking some tiny amount of current through the dielectric, from one terminal to the other. This tiny current loss (usually nanoamps or less) is called leakage. Leakage causes energy stored in the capacitor to slowly, but surely drain away.
- Equivalent series resistance (ESR) - The terminals of a capacitor aren't 100% conductive, they'll always have a tiny amount of resistance (usually less than 0.01Ω) to them. This resistance becomes a problem when a lot of current runs through the cap, producing heat and power loss.
- Tolerance - Capacitors also can't be made to have an exact, precise capacitance. Each cap will be rated for their nominal capacitance, but, depending on the type, the exact value might vary anywhere from ±1% to ±20% of the desired value.

Ceramic Capacitors

The most commonly used and produced capacitor out there is the ceramic capacitor. The name comes from the material from which their dielectric is made.

Ceramic capacitors are usually both physically and capacitance-wise small. It's hard to find a ceramic capacitor much larger than 10µF. A surface-mount ceramic cap is commonly found in a tiny 0402 (0.4mm x 0.2mm), 0603 (0.6mm x 0.3mm) or 0805 package. Through-hole ceramic caps usually look like small (commonly yellow or red) bulbs, with two protruding terminals.

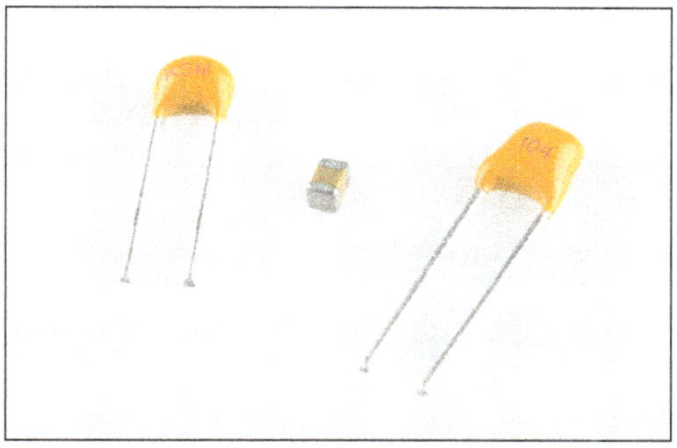

Two caps in a through-hole, radial package; a 22pF cap on the left, and a 0.1µF on the right. In the middle, a tiny 0.1µF 0603 surface-mount cap.

Compared to the equally popular electrolytic caps, ceramics are a more near-ideal capacitor (much lower ESR and leakage currents), but their small capacitance can be limiting. They are usually the least expensive option too. These caps are well-suited for high-frequency coupling and decoupling applications.

Aluminum and Tantalum Electrolytic

Electrolytes are great because they can pack *a lot* of capacitance into a relatively small volume. If you need a capacitor in the range of 1µF-1mF, you're most likely to find it in an electrolytic form. They're especially well-suited to high-voltage applications because of their relatively high maximum voltage ratings.

Aluminum electrolytic capacitors, the most popular of the electrolytic family, usually look like little tin cans, with both leads extending from the bottom.

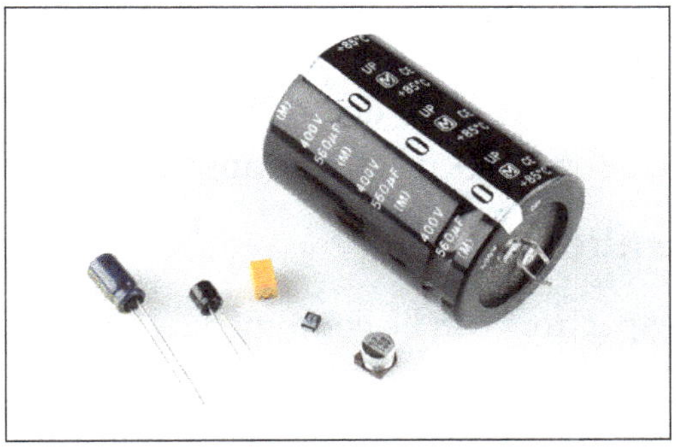

An assortment of through-hole and surface-mount electrolytic capacitors

Unfortunately, electrolytic caps are usually polarized. They have a positive pin - the anode - and a negative pin called the cathode. When voltage is applied to an electrolytic cap, the anode must be at a higher voltage than the cathode. The cathode of an electrolytic capacitor is usually identified with a marking, and a colored strip on the case. The leg of the anode might also be slightly longer as another indication. If voltage is applied in reverse on an electrolytic cap, they'll fail spectacularly (making a *pop* and bursting open), and permanently. After popping an electrolytic will behave like a short circuit.

These caps also notorious for leakage - allowing small amounts of current (on the order of nA) to run through the dielectric from one terminal to the other. This makes electrolytic caps less-than-ideal for energy storage, which is unfortunate given their high capacity and voltage rating.

Super Capacitors

These caps are uniquely designed to have *very* high capacitances, in the range of farads.

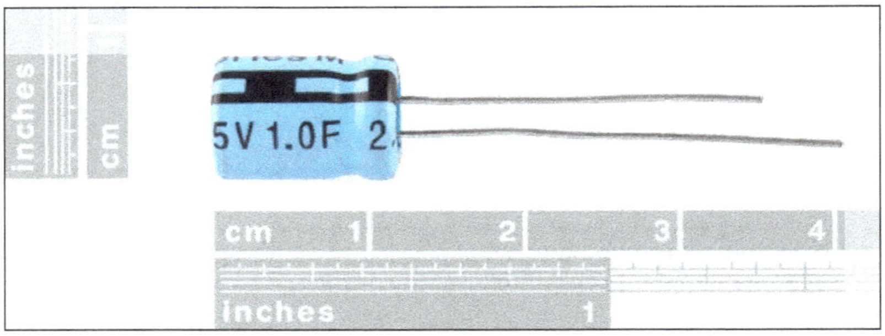

Super capacitor: High capacitance, but only rated for 2.5V.

While they can store a huge amount of charge, super caps can't deal with very high voltages. This 10F super cap is only rated for 2.5V max. Any more than that will destroy it. Super caps are commonly placed in series to achieve a higher voltage rating (while reducing total capacitance).

The main application for super capacitors is in storing and releasing energy, like batteries, which are their main competition. While super caps can't hold as much energy as an equally sized battery, they can release it much faster, and they usually have a much longer lifespan.

Electrolytic and ceramic caps cover about 80% of the capacitor types out there. Another common capacitor type is the film capacitor, which features very low parasitic losses (ESR), making them great for dealing with very high currents.

There's plenty of other less common capacitors. Variable capacitors can produce a range of capacitances, which makes them a good alternative to variable resistors in tuning circuits. Twisted wires or PCBs can create capacitance (sometimes undesired) because each consists of two conductors separated by an insulator. Leyden Jars -- a glass jar filled with and surrounded by conductors - are the O.G. of the capacitor family. Flux capacitors (a strange combination of inductor and capacitor.

Capacitors in Series/Parallel

Much like resistors, multiple capacitors can be combined in series or parallel to create a combined equivalent capacitance. Capacitors, however, add together in a way that's completely the opposite of resistors.

Capacitors in Parallel

When capacitors are placed in parallel with one another the total capacitance is simply the sum of all capacitances. This is analogous to the way resistors add when in series.

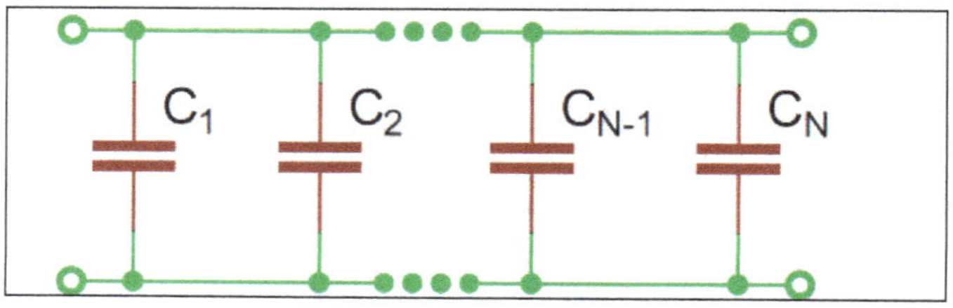

$$C_{Tot} = C_1 + C_2 ... + C_{N-1} + C_N$$

So, for example, if you had three capacitors of values 10µF, 1µF, and 0.1µF in parallel, the total capacitance would be 11.1µF (10+1+0.1).

Capacitors in Series

Much like resistors are a pain to add in parallel, capacitors get funky when placed in *series*. The total capacitance of *N* capacitors in series is the inverse of the sum of all inverse capacitances.

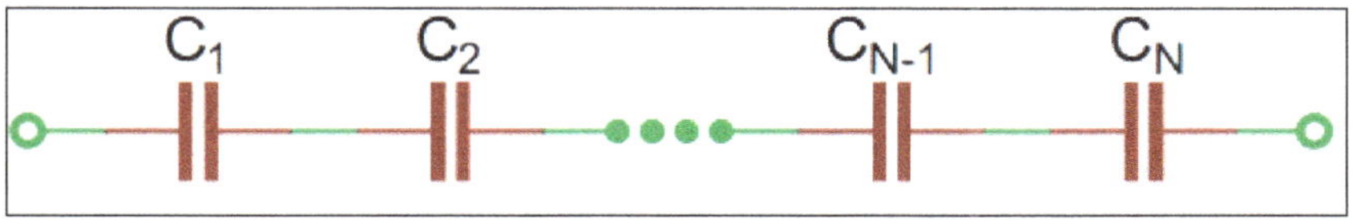

$$\frac{1}{C_{Tot}} = \frac{1}{C_1} + \frac{1}{C_2} + ... + \frac{1}{C_{N-1}} + \frac{1}{C_N}$$

If you only have two capacitors in series, you can use the "product-over-sum" method to calculate the total capacitance:

$$C_{tot} = \frac{C_1 . C_2}{C_1 + C_2}$$

Taking that equation even further, if you have two equal-valued capacitors in series, the total capacitance is half of their value. For example two 10F super capacitors in series will produce a total capacitance of 5F (it'll also have the benefit of doubling the voltage rating of the total capacitor, from 2.5V to 5V).

Application Examples

There are tons of applications for this nifty little (actually they're usually pretty large) passive component. To give you an idea of their wide range of uses, here are a few examples:

Decoupling (Bypass) Capacitors

A lot of the capacitors you see in circuits, especially those featuring an integrated circuit, are decoupling. A decoupling capacitor's job is to supress high-frequency noise in power supply signals. They take tiny voltage ripples, which could otherwise be harmful to delicate ICs, out of the voltage supply.

In a way, decoupling capacitors act as a very small, local power supply for ICs (almost like an uninterruptible power supply is to computers). If the power supply very temporarily drops its voltage (which is actually pretty common, especially when the circuit it's powering is constantly switching its load requirements), a decoupling capacitor can briefly supply power at the correct voltage. This is why these capacitors are also called bypass caps; they can temporarily act as a power source, *bypassing* the power supply.

Decoupling capacitors connect between the power source (5V, 3.3V, etc.) and ground. It's not uncommon to use two or more different-valued, even different types of capacitors to bypass the power supply, because some capacitor values will be better than others at filtering out certain frequencies of noise.

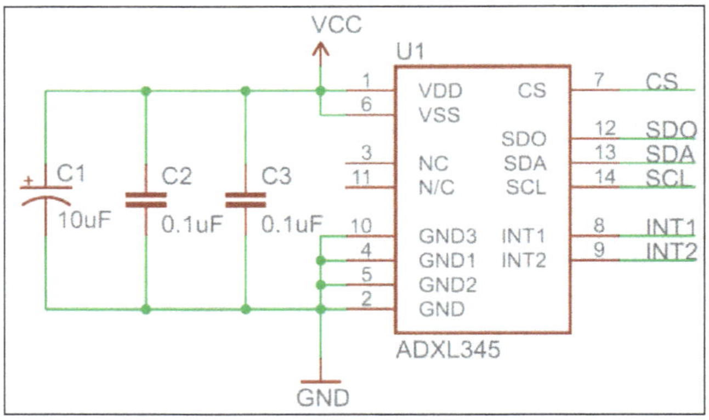

In this schematic, three decoupling capacitors are used to help reduce the noise in an accelerometer's voltage supply. Two ceramic 0.1μF and one tantalum electrolytic 10μF split decoupling duties.

While it seems like this might create a short from power to ground, only high-frequency signals can run through the capacitor to ground. The DC signal will go to the IC, just as desired. Another reason these are called bypass capacitors is because the high frequencies (in the kHz-MHz range) bypass the IC, instead running through the capacitor to get to ground.

When physically placing decoupling capacitors, they should always be located as close as possible to an IC. The further away they are, they less effective they'll be.

The tiny, black IC is surrounded by two 0.1μF capacitors (the brown caps) and one 10μF electrolytic tantalum capacitor (the tall, black/grey rectangular cap).

To follow good engineering practice, always add at least one decoupling capacitor to every IC. Usually 0.1μF is a good choice, or even add some 1μF or 10μF caps. They're a cheap addition, and they help make sure the chip isn't subjected to big dips or spikes in voltage.

Power Supply Filtering

Diode rectifiers can be used to turn the AC voltage coming out of your wall into the DC voltage required by most electronics. But diodes alone can't turn an AC signal into a clean DC signal, they need the help of capacitors! By adding a parallel capacitor to a bridge rectifier, a rectified signal like this,

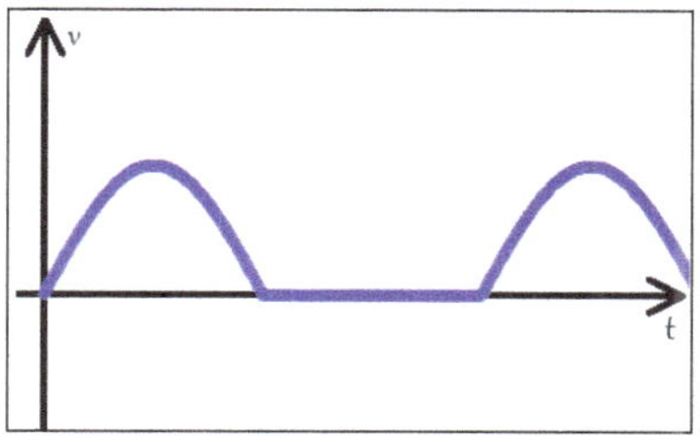

Can be turned into a near-level DC signal like this,

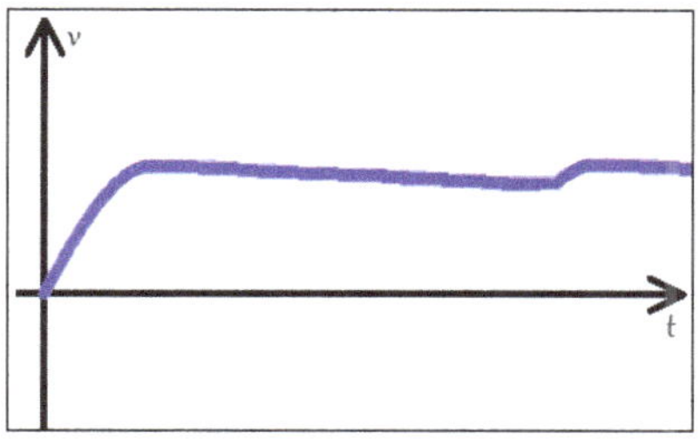

Capacitors are stubborn components; they'll always try to resist sudden changes in voltage. The filter capacitor will charge up as the rectified voltage increases. When the rectified voltage coming into the cap starts its rapid decline, the capacitor will access its bank of stored energy, and it'll discharge very slowly, supplying energy to the load. The capacitor shouldn't fully discharge before the input rectified signal starts to increase again, recharging the cap. This dance plays out many times a second, over-and-over as long as the power supply is in use.

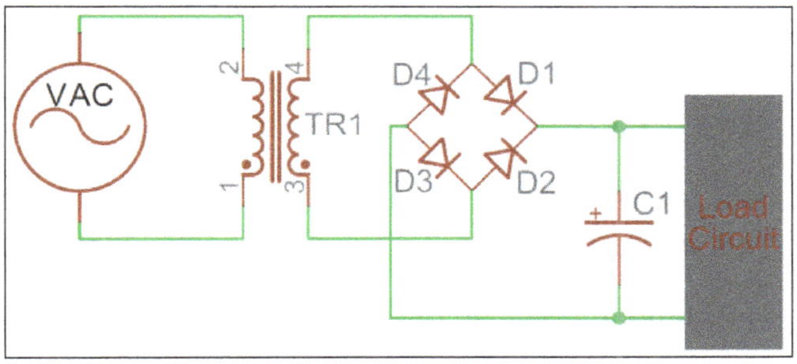

An AC-to-DC power supply circuit. The filter cap (C1) is critical in smoothing out the DC signal sent to the load circuit.

If you tear apart any AC-to-DC power supply, you're bound to find at least one rather large capacitor. Below are the guts of a 9V DC wall adapter.

There might be more capacitors than you think. There are four electrolytic, tin-can-looking caps ranging from 47μF to 1000μF. The big, yellow rectangle in the foreground is a high-voltage 0.1μF polypropylene film cap. The blue disc-shaped cap and the little green one in the middle are both ceramics.

Energy Storage and Supply

It seems obvious that if a capacitor stores energy, one of it's many applications would be supplying that energy to a circuit, just like a battery. The problem is capacitors have a much lower energy density than batteries; they just can't pack as much energy as an equally sized chemical battery.

The upside of capacitors is they usually lead longer lives than batteries, which makes them a better choice environmentally. They're also capable of delivering energy much faster than a battery, which makes them good for applications which need a short, but high burst of power. A camera flash might get its power from a capacitor (which, in turn, was probably charged by a battery).

Battery or Capacitor

	Battery	Capacitor
Capacity	✓	
Energy Density	✓	
Charge/Discharge Rate		✓
Life Span		✓

Signal Filtering

Capacitors have a unique response to signals of varying frequencies. They can block out low-frequency or DC signal-components while allowing higher frequencies to pass right through. They're like a bouncer at a very exclusive club for high frequencies only.

Filtering signals can be useful in all sorts of signal processing applications. Radio receivers might use a capacitor (among other components) to tune out undesired frequencies.

Another example of capacitor signal filtering is passive crossover circuits inside speakers, which separate a single audio signal into many. A series capacitor will block out low frequencies, so the remaining high-frequency parts of the signal can go to the speaker's tweeter. In the low-frequency passing, subwoofer circuit, high-frequencies can mostly be shunted to ground through the parallel capacitor.

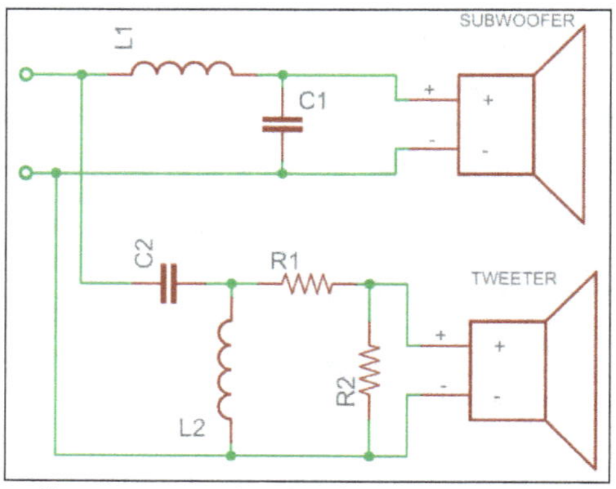

A very simple example of an audio crossover circuit. The capacitor will block out low frequencies, while the inductor blocks out high frequencies. Each can be used to deliver the proper signal to tuned audio drivers.

Types of Capacitor

The types of capacitors available range from very small delicate trimming capacitors using in oscillator or radio circuits, up to large power metal-can type capacitors used in high voltage power correction and smoothing circuits.

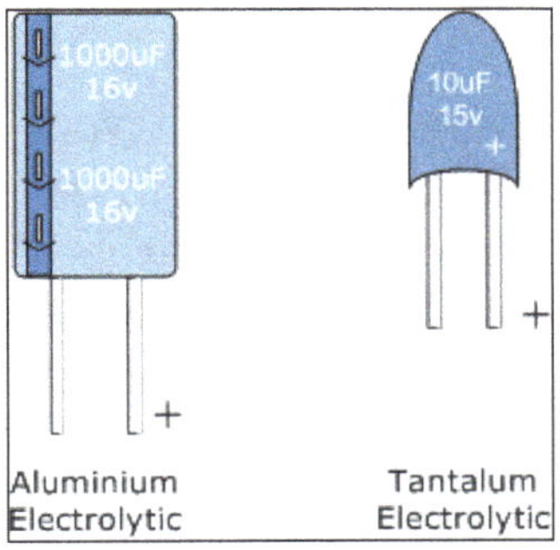

The comparisons between the different *types of capacitor* is generally made with regards to the dielectric used between the plates. Like resistors, there are also variable types of capacitors which allow us to vary their capacitance value for use in radio or "frequency tuning" type circuits.

Commercial types of capacitors are made from metallic foil interlaced with thin sheets of either paraffin-impregnated paper or Mylar as the dielectric material. Some capacitors look like tubes, this is because the metal foil plates are rolled up into a cylinder to form a small package with the insulating dielectric material sandwiched in between them.

Small capacitors are often constructed from ceramic materials and then dipped into an epoxy resin to seal them. Either way, capacitors play an important part in electronic circuits so here are a few of the more "common" types of capacitor available.

Dielectric Capacitor

Dielectric Capacitors are usually of the variable type were a continuous variation of capacitance is required for tuning transmitters, receivers and transistor radios. Variable dielectric capacitors are multi-plate air-spaced types that have a set of fixed plates (the stator vanes) and a set of movable plates (the rotor vanes) which move in between the fixed plates.

The position of the moving plates with respect to the fixed plates determines the overall capacitance value. The capacitance is generally at maximum when the two sets of plates are fully meshed together. High voltage type tuning capacitors have relatively large spacing's or air-gaps between the plates with breakdown voltages reaching many thousands of volts.

Variable Capacitor Symbol

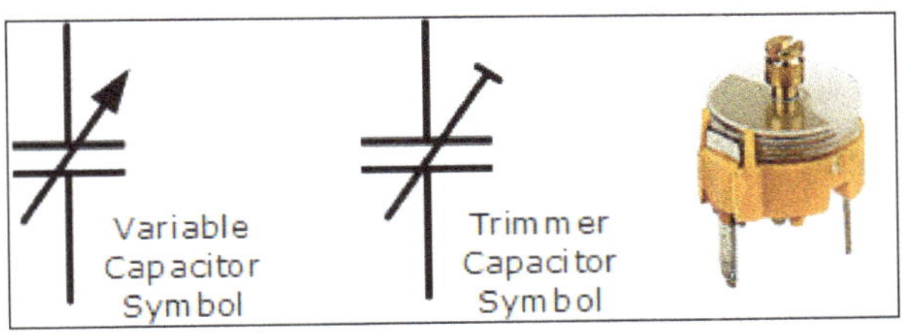

As well as the continuously variable types, preset type variable capacitors are also available called Trimmers. These are generally small devices that can be adjusted or "pre-set" to a particular capacitance value with the aid of a small screwdriver and are available in very small capacitance's of 500pF or less and are non-polarized.

Film Capacitor Type

Film Capacitors are the most commonly available of all types of capacitors, consisting of a relatively large family of capacitors with the difference being in their dielectric properties. These include polyester (Mylar), polystyrene, polypropylene, polycarbonate, metalised paper, Teflon etc. Film type capacitors are available in capacitance ranges from as small as 5pF to as large as 100uFdepending upon the actual type of capacitor and its voltage rating. Film capacitors also come in an assortment of shapes and case styles which include:

- Wrap & Fill (Oval & Round) – where the capacitor is wrapped in a tight plastic tape and have the ends filled with epoxy to seal them.
- Epoxy Case (Rectangular & Round) – where the capacitor is encased in a moulded plastic shell which is then filled with epoxy.
- Metal Hermetically Sealed (Rectangular & Round) – where the capacitor is encased in a metal tube or can and again sealed with epoxy.

Film Capacitors which use polystyrene, polycarbonate or Teflon as their dielectrics are sometimes called "Plastic capacitors". The construction of plastic film capacitors is similar to that for paper film capacitors but use a plastic film instead of paper. The main advantage of plastic film capacitors compared to impregnated-paper types is that they operate well under conditions of high temperature, have smaller tolerances, a very long service life and high reliability. Examples of film capacitors are the rectangular metalised film and cylindrical film & foil types as shown below.

Radial Lead Type

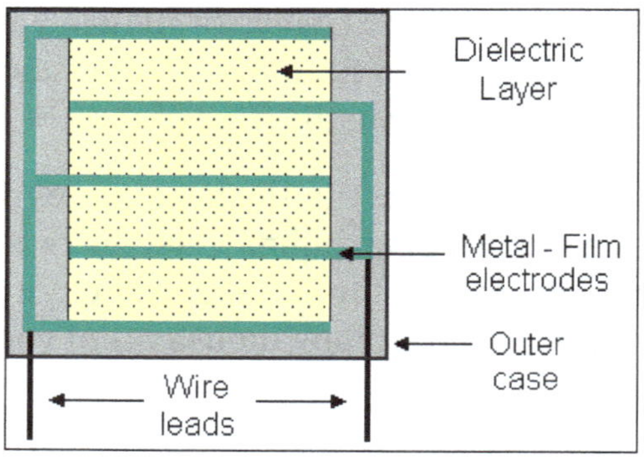

Axial Lead Type

The film and foil types of capacitors are made from long thin strips of thin metal foil with the dielectric material sandwiched together which are wound into a tight roll and then sealed in paper or metal tubes.

These film types require a much thicker dielectric film to reduce the risk of tears or punctures in the film, and is therefore more suited to lower capacitance values and larger case sizes.

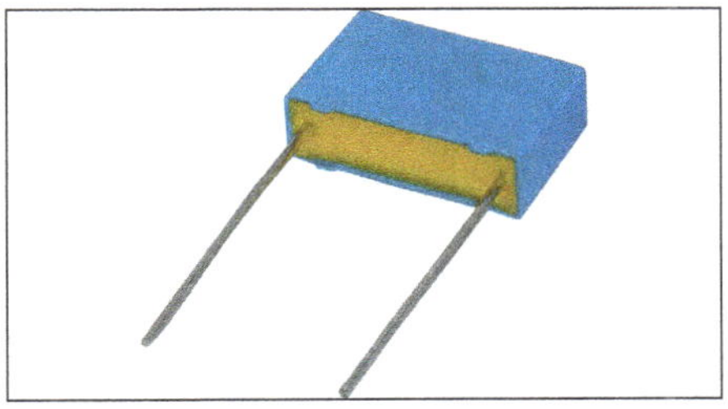

Film Capacitor

Metalised foil capacitors have the conductive film metalised sprayed directly onto each side of the dielectric which gives the capacitor self-healing properties and can therefore use much thinner dielectric films. This allows for higher capacitance values and smaller case sizes for a given capacitance. Film and foil capacitors are generally used for higher power and more precise applications.

Ceramic Capacitors

Ceramic Capacitors or Disc Capacitors as they are generally called are made by coating two sides of a small porcelain or ceramic disc with silver and are then stacked together to make a capacitor. For very low capacitance values a single ceramic disc of about 3-6mm is used. Ceramic capacitors have a high dielectric constant (High-K) and are available so that relatively high capacitance's can be obtained in a small physical size.

Ceramic Capacitor

They exhibit large non-linear changes in capacitance against temperature and as a result are used as de-coupling or by-pass capacitors as they are also non-polarized devices. Ceramic capacitors have values ranging from a few pico-farads to one or two microfarads, (μF) but their voltage ratings are generally quite low.

Ceramic types of capacitors generally have a 3-digit code printed onto their body to identify their capacitance value in pico-farads. Generally the first two digits indicate the capacitors value and the third digit indicates the number of zero's to be added. For example, a ceramic disc capacitor with the markings 103 would indicate 10 and 3 zero's in pico-farads which is equivalent to 10,000 pF or 10nF.

Likewise, the digits 104 would indicate 10 and 4 zero's in pico-farads which is equivalent to 100,000 p For 100nF and so on. So on the image of the ceramic capacitor above the numbers 154 indicate 15 and 4 zero's in pico-farads which is equivalent to 150,000 pF or 150nF or 0.15μF. Letter codes are sometimes used to indicate their tolerance value such as: J = 5%, K = 10% or M = 20% etc.

Electrolytic Capacitors

Electrolytic Capacitors are generally used when very large capacitance values are required. Here instead of using a very thin metallic film layer for one of the electrodes, a semi-liquid electrolyte solution in the form of a jelly or paste is used which serves as the second electrode (usually the cathode).

The dielectric is a very thin layer of oxide which is grown electro-chemically in production with the thickness of the film being less than ten microns. This insulating layer is so thin that it is possible to make capacitors with a large value of capacitance for a small physical size as the distance between the plates, d is very small.

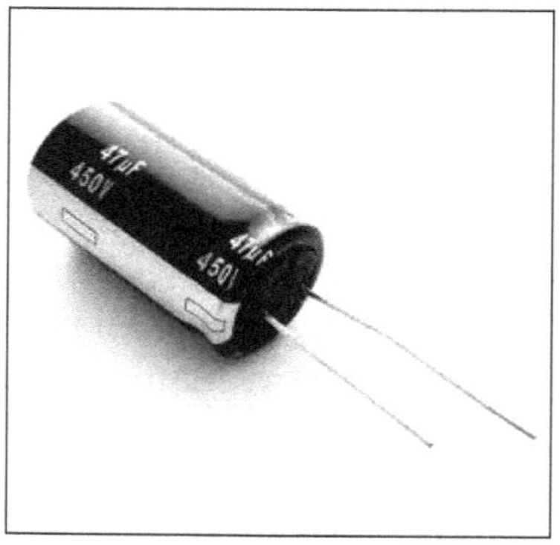

Electrolytic Capacitor

The majority of electrolytic types of capacitors are Polarised, that is the DC voltage applied to the capacitor terminals must be of the correct polarity, i.e. positive to the positive terminal and negative to the negative terminal as an incorrect polarisation will break down the insulating oxide layer and permanent damage may result.

All polarised electrolytic capacitors have their polarity clearly marked with a negative sign to indicate the negative terminal and this polarity must be followed.

Electrolytic Capacitors are generally used in DC power supply circuits due to their large capacitance's and small size to help reduce the ripple voltage or for coupling and decoupling applications. One main disadvantage of electrolytic capacitors is their relatively low voltage rating and due to the polarisation of electrolytic capacitors, it follows then that they must not be used on AC supplies. Electrolytic's generally come in two basic forms; Aluminium Electrolytic Capacitors and Tantalum Electrolytic Capacitors.

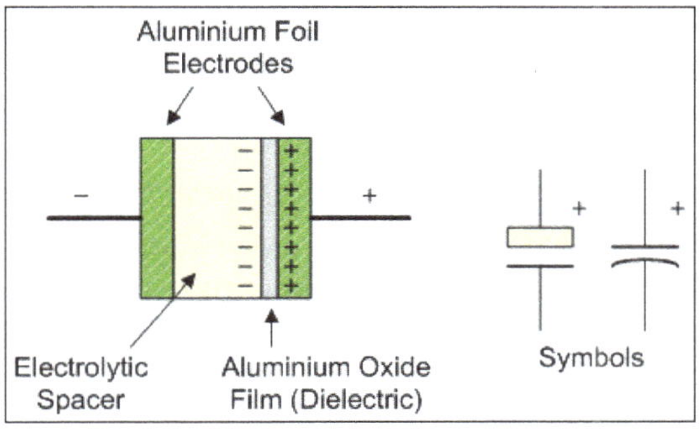

Aluminium Electrolytic Capacitors

There are basically two types of Aluminium Electrolytic Capacitor, the plain foil type and the etched foil type. The thickness of the aluminium oxide film and high breakdown voltage give these capacitors very high capacitance values for their size.

The foil plates of the capacitor are anodized with a DC current. This anodizing process sets up the polarity of the plate material and determines which side of the plate is positive and which side is negative.

The etched foil type differs from the plain foil type in that the aluminium oxide on the anode and cathode foils has been chemically etched to increase its surface area and permittivity. This gives a smaller sized capacitor than a plain foil type of equivalent value but has the disadvantage of not being able to withstand high DC currents compared to the plain type. Also their tolerance range is quite large at up to 20%. Typical values of capacitance for an aluminium electrolytic capacitor range from 1uF up to 47,000uF.

Etched foil electrolytic's are best used in coupling, DC blocking and by-pass circuits while plain foil types are better suited as smoothing capacitors in power supplies. But aluminium electrolytic's are "polarised" devices so reversing the applied voltage on the leads will cause the insulating layer within the capacitor to become destroyed along with the capacitor. However, the electrolyte used within the capacitor helps heal a damaged plate if the damage is small.

Since the electrolyte has the properties to self-heal a damaged plate, it also has the ability to re-anodize the foil plate. As the anodizing process can be reversed, the electrolyte has the ability to remove the oxide coating from the foil as would happen if the capacitor was connected with a reverse polarity. Since the electrolyte has the ability to conduct electricity, if the aluminium oxide layer was removed or destroyed, the capacitor would allow current to pass from one plate to the other destroying the capacitor.

Tantalum Electrolytic Capacitors

Tantalum Electrolytic Capacitors and Tantalum Beads are available in both wet (foil) and dry (solid) electrolytic types with the dry or solid tantalum being the most common. Solid tantalum capacitors use manganese dioxide as their second terminal and are physically smaller than the equivalent aluminium capacitors.

The dielectric properties of tantalum oxide is also much better than those of aluminium oxide giving a lower leakage currents and better capacitance stability which makes them suitable for use in blocking, by-passing, decoupling, filtering and timing applications.

Also, Tantalum Capacitors although polarised, can tolerate being connected to a reverse voltage much more easily than the aluminium types but are rated at much lower working voltages. Solid tantalum capacitors are usually used in circuits where the AC voltage is small compared to the DC voltage.

However, some tantalum capacitor types contain two capacitors in-one, connected negative-to-negative to form a "non-polarised" capacitor for use in low voltage AC circuits as a non-polarised device. Generally, the positive lead is identified on the capacitor body by a polarity mark, with the body of a tantalum bead capacitor being an oval geometrical shape. Typical values of capacitance range from 47nF to 470uF.

Aluminium and Tantalum Electrolytic Capacitor

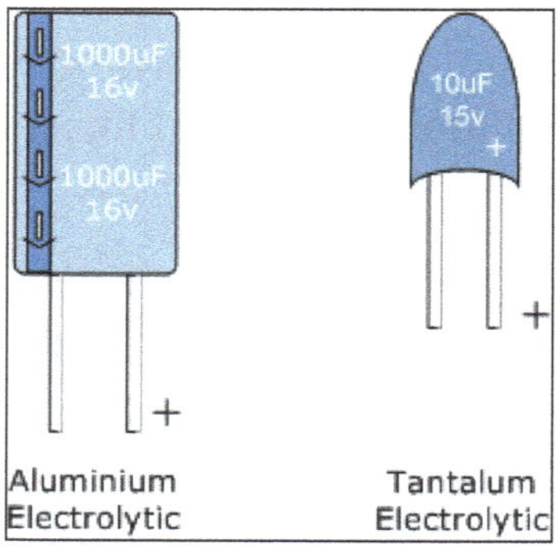

Electrolytic's are widely used capacitors due to their low cost and small size but there are three easy ways to destroy an electrolytic capacitor:

- Over-voltage – excessive voltage will cause current to leak through the dielectric resulting in a short circuit condition.
- Reversed Polarity – reverse voltage will cause self-destruction of the oxide layer and failure.
- Over Temperature – excessive heat dries out the electrolytic and shortens the life of an electrolytic capacitor.

Capacitor Characteristics

The characteristics of capacitors define its temperature, voltage rating and capacitance range as well as its use in a particular application.

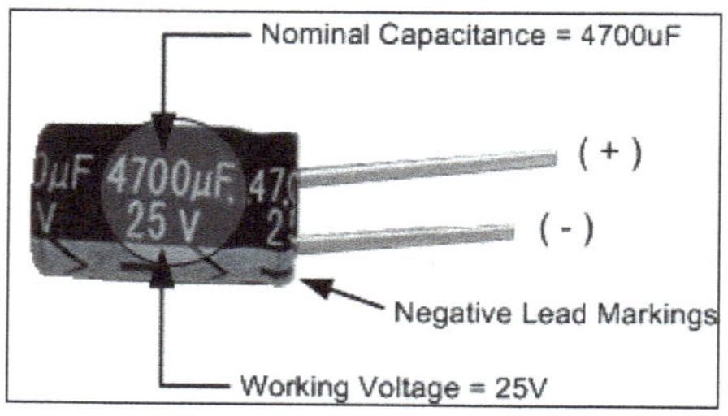

The capacitor, as with any other electronic component, comes defined by a series of characteristics. These Capacitor Characteristics can always be found in the data sheets that the capacitor manufacturer provides to us.

Nominal Capacitance (C)

The nominal value of the Capacitance, C of a capacitor is the most important of all capacitor characteristics. This value measured in pico-Farads (pF), nano-Farads (nF) or micro-Farads (μF) and is marked onto the body of the capacitor as numbers, letters or coloured bands.

The capacitance of a capacitor can change value with the circuit frequency (Hz) y with the ambient temperature. Smaller ceramic capacitors can have a nominal value as low as one pico-Farad, (1pF) while larger electrolytic's can have a nominal capacitance value of up to one Farad, (1F).

All capacitors have a tolerance rating that can range from -20% to as high as +80% for aluminium electrolytic's affecting its actual or real value. The choice of capacitance is determined by the circuit configuration but the value read on the side of a capacitor may not necessarily be its actual value.

Working Voltage (WV)

The Working Voltage is another important capacitor characteristic that defines the maximum continuous voltage either DC or AC that can be applied to the capacitor without failure during its working life. Generally, the working voltage printed onto the side of a capacitors body refers to its DC working voltage, (WVDC).

DC and AC voltage values are usually not the same for a capacitor as the AC voltage value refers to the r.m.s. value and NOT the maximum or peak value which is 1.414 times greater. Also, the specified DC working voltage is valid within a certain temperature range, normally -30 °C to +70 °C.

Any DC voltage in excess of its working voltage or an excessive AC ripple current may cause failure. It follows therefore, that a capacitor will have a longer working life if operated in a cool environment and within its rated voltage. Common working DC voltages are 10V, 16V, 25V, 35V, 50V, 63V, 100V, 160V, 250V, 400V and 1000V and are printed onto the body of the capacitor.

Tolerance (±%)

As with resistors, capacitors also have a Tolerance rating expressed as a plus-or-minus value either in picofarad's (±pF) for low value capacitors generally less than 100pF or as a percentage (±%) for higher value capacitors generally higher than 100pF.

The tolerance value is the extent to which the actual capacitance is allowed to vary from its nominal value and can range anywhere from -20% to +80%. Thus a 100μF capacitor with a ±20% tolerance could legitimately vary from 80μF to 120μF and still remain within tolerance.

Capacitors are rated according to how near to their actual values they are compared to the rated nominal capacitance with coloured bands or letters used to indicated their actual tolerance. The most common tolerance variation for capacitors is 5% or 10% but some plastic capacitors are rated as low as ±1%.

Leakage Current

The dielectric used inside the capacitor to separate the conductive plates is not a perfect insulator resulting in a very small current flowing or "leaking" through the dielectric due to the influence of the powerful electric fields built up by the charge on the plates when applied to a constant supply voltage.

This small DC current flow in the region of nano-amps (nA) is called the capacitors Leakage Current. Leakage current is a result of electrons physically making their way through the dielectric medium, around its edges or across its leads and which will over time fully discharging the capacitor if the supply voltage is removed.

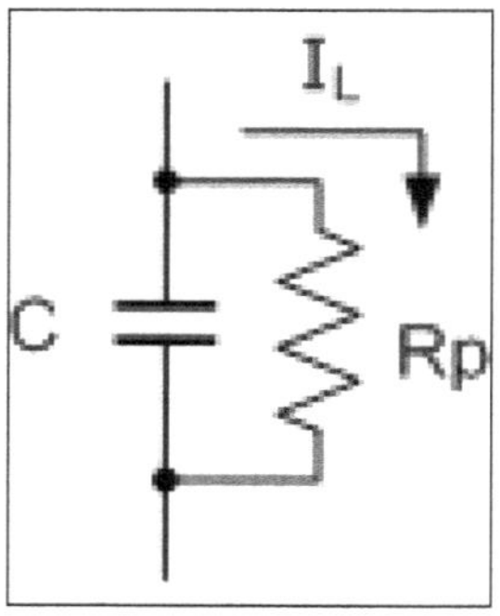

Leakage Model

When the leakage is very low such as in film or foil type capacitors it is generally referred to as "insulation resistance" (R_p) and can be expressed as a high value resistance in parallel with the capacitor as shown. When the leakage current is high as in electrolytic's it is referred to as a "leakage current" as electrons flow directly through the electrolyte.

Capacitor leakage current is an important parameter in amplifier coupling circuits or in power supply circuits, with the best choices for coupling and storage applications being Teflon and the other plastic capacitor types (polypropylene, polystyrene, etc) because the lower the dielectric constant, the higher the insulation resistance.

Electrolytic-type capacitors (tantalum and aluminium) on the other hand may have very high capacitances, but they also have very high leakage currents (typically of the order of about 5-20 μA per μF) due to their poor isolation resistance, and are therefore not suited for storage or coupling applications. Also, the flow of leakage current for aluminium electrolytic's increases with temperature.

Working Temperature (T)

Changes in temperature around the capacitor affect the value of the capacitance because of changes in the dielectric properties. If the air or surrounding temperature becomes hot or cold the capacitance value of the capacitor may change so much as to affect the correct operation of the circuit. The normal working range for most capacitors is -30 °C to +125 °C with nominal voltage ratings given for a Working Temperature of no more than +70 °C especially for the plastic capacitor types.

Generally for electrolytic capacitors and especially aluminium electrolytic capacitor, at high temperatures (over +85 °C the liquids within the electrolyte can be lost to evaporation, and the body of the capacitor (especially the small sizes) may become deformed due to the internal pressure and leak outright. Also, electrolytic capacitors can not be used at low temperatures, below about -10 °C, as the electrolyte jelly freezes.

Temperature Coefficient (TC)

The Temperature Coefficient of a capacitor is the maximum change in its capacitance over a specified temperature range. The temperature coefficient of a capacitor is generally expressed linearly as parts per million per degree centigrade (PPM/°C), or as a percent change over a particular range of temperatures. Some capacitors are non linear (Class 2 capacitors) and increase their value as the temperature rises giving them a temperature coefficient that is expressed as a positive "P".

Some capacitors decrease their value as the temperature rises giving them a temperature coefficient that is expressed as a negative "N". For example "P100" is +100 ppm/°C or "N200", which is -200 ppm/°C etc. However, some capacitors do not change their value and remain constant over a certain temperature range, such capacitors have a zero temperature coefficient or "NPO". These types of capacitors such as Mica or Polyester are generally referred to as Class 1 capacitors.

Most capacitors, especially electrolytic's lose their capacitance when they get hot but temperature compensating capacitors are available in the range of at least P1000 through to N5000 (+1000 ppm/°C through to -5000 ppm/°C). It is also possible to connect a capacitor with a positive temperature coefficient in series or parallel with a capacitor having a negative temperature coefficient the net result being that the two opposite effects will cancel each other out over a certain range of temperatures. Another useful application of temperature coefficient capacitors is to use them to cancel out the effect of temperature on other components within a circuit, such as inductors or resistors etc.

Polarization

Capacitor Polarization generally refers to the electrolytic type capacitors but mainly the Aluminium Electrolytic's, with regards to their electrical connection. The majority of electrolytic capacitors are polarized types that is the voltage connected to the capacitor terminals must have the correct polarity, i.e. positive to positive and negative to negative.

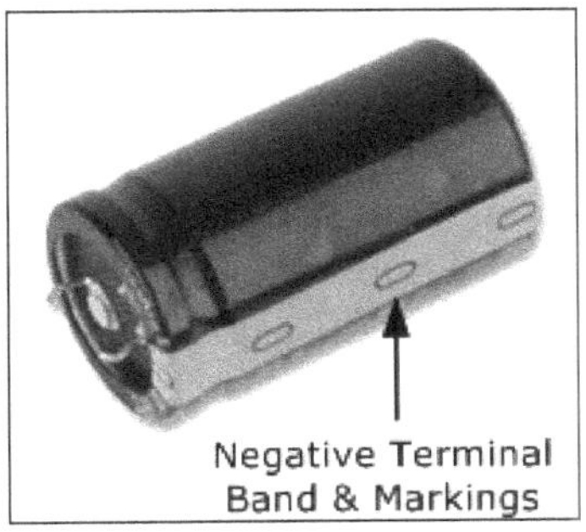

Incorrect polarization can cause the oxide layer inside the capacitor to break down resulting in very large currents flowing through the device resulting in destruction .

The majority of electrolytic capacitors have their negative, -ve terminal clearly marked with either a black stripe, band, arrows or chevrons down one side of their body as shown, to prevent any incorrect connection to the DC supply.

Some larger electrolytic's have their metal can or body connected to the negative terminal but high voltage types have their metal can insulated with the electrodes being brought out to separate spade or screw terminals for safety.

Also, when using aluminium electrolytic's in power supply smoothing circuits care should be taken to prevent the sum of the peak DC voltage and AC ripple voltage from becoming a "reverse voltage".

Equivalent Series Resistance (ESR)

The Equivalent Series Resistance or ESR, of a capacitor is the AC impedance of the capacitor when used at high frequencies and includes the resistance of the dielectric material, the DC resistance of the terminal leads, the DC resistance of the connections to the dielectric and the capacitor plate resistance all measured at a particular frequency and temperature.

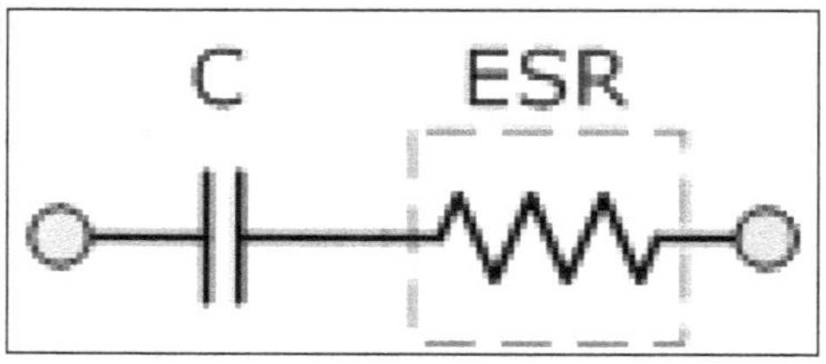

ESR Model

In some ways, ESR is the opposite of the insulation resistance which is presented as a pure resistance (no capacitive or inductive reactance) in parallel with the capacitor. An ideal capacitor would have only capacitance but ESR is presented as a pure resistance (less than 0.1Ω) in series with the capacitor (hence the name Equivalent Series Resistance), and which is frequency dependent making it a "DYNAMIC" quantity.

As ESR defines the energy losses of the "equivalent" series resistance of a capacitor it must therefore determine the capacitor's overall I^2R heating losses especially when used in power and switching circuits.

Capacitors with a relatively high ESR have less ability to pass current to and from its plates to the external circuit because of their longer charging and discharging RC time constant. The ESR of electrolytic capacitors increases over time as their electrolyte dries out. Capacitors with very low ESR ratings are available and are best suited when using the capacitor as a filter.

Capacitors with small capacitance's (less than 0.01μF) generally do not pose much danger to humans. However, when their capacitance's start to exceed 0.1μF, touching the capacitor leads can be a shocking experience.

Capacitors have the ability to store an electrical charge in the form of a voltage across themselves even when there is no circuit current flowing, giving them a sort of memory with large electrolytic type reservoir capacitors found in television sets, photo flashes and capacitor banks potentially storing a lethal charge.

Never touch the leads of large value capacitors once the power supply is removed. If you are unsure about their condition or the safe handling of these large capacitors, seek help or expert advice before handling them.

Inductor

The inductor is perhaps the simplest of all electronic components, constructed much like a resistor – a simple length of wire that is coiled up. However, here, resistance is not the property we're looking for. It is something that happens because of the shape of the wire – a coil – it creates a magnetic field when a current is passed through it. This induced magnetic field gives this bit of wire some interesting electrical properties, especially inductance – which gives these parts their name.

Symbols for Inductors

Like many other electronic components, the symbol for an inductor is a simplified pictogram of what it actually looks like:

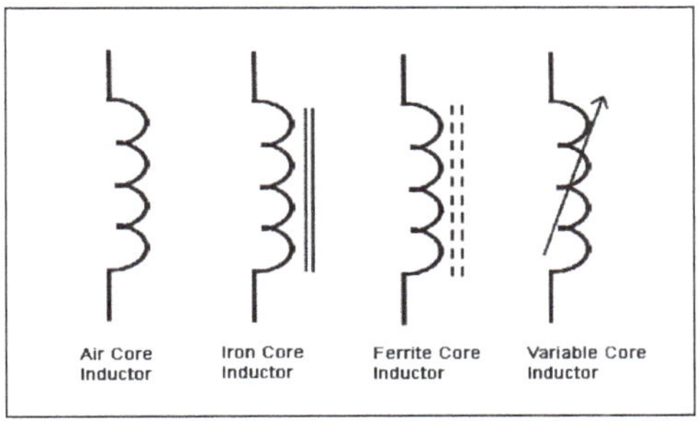

The lines near the symbol represent the core material.

Working of an Inductor

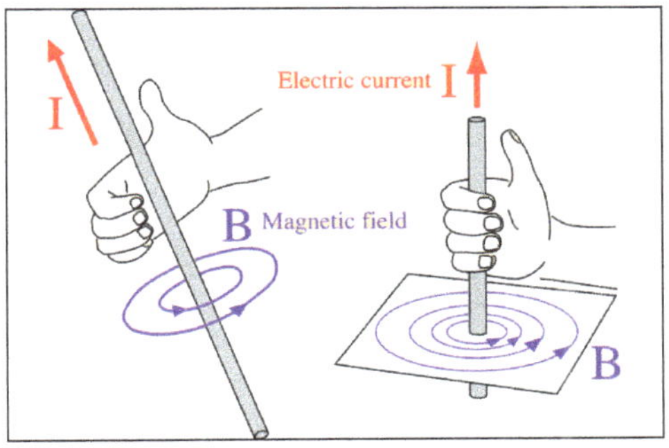

However, if you plug in the value of the current into the formulas, then you'll realize that the magnetic field produced is tiny – almost negligible, unless the currents are impossibly high, in the order of mega-amps.

So in order to increase the magnetic field created by a specific length of wire, we wrap it into the form of a coil. This increases the magnetic field, like so:

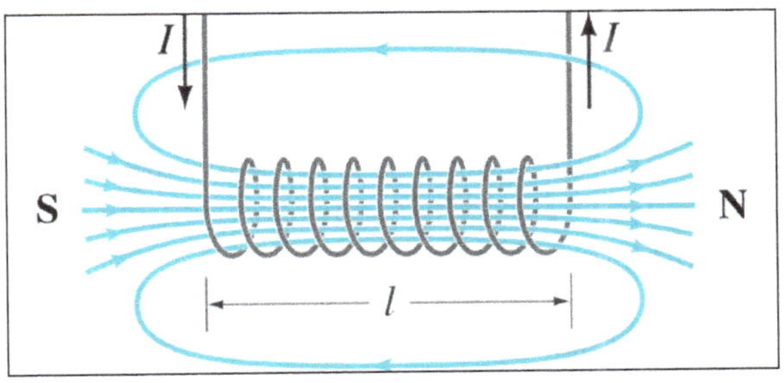

This shape is also called a solenoid.

When a voltage is applied across the terminals of an inductor, the current flowing creates a magnetic field. This magnetic field again creates an induced current in the inductor of opposite polarity, according to Lenz's law. The currents do not cancel each other out – rather, the induced current actively tries to oppose the incoming current due to the voltage across the inductor. The overall result of this battle is that the current through an inductor cannot change rapidly – it is always a linear slope.

Measuring an Inductor

We could use the same approach we used for capacitors, and we can define inductance of a circuit as the voltage change induced when the current changes at a certain rate.

We could try measuring inductors by the magnetic field that they create. As soon as we do that, we run into problems. The magnetic field created by an inductor depends on the current that passes through it, so even a small inductor can create a large magnetic field.

Instead, we could use the same approach we used for capacitors, and we can define inductance of a circuit as the voltage change induced when the current changes at a certain rate.

Mathematically,

$V = L(dI/dt)$

Where V is the voltage, L is the inductance, I is the current and t is the time period.

Inductance, 'L', is measured in Henrys, named after Joseph Henry, the American scientist who discovered electromagnetic induction.

The formula for calculating the inductance of a coil of wire is given by this formula:

$L = (\mu n2a)/l$

Where L is the inductance in Henrys, μ is the permeability constant, i.e. a coefficient of how easily the magnetic field can be created in a given medium, n is the number of turns, a is the area of the coil and l is the length of the coil.

Again, the Henry is a very large unit, so practically inductors are measured in microHenrys, uH, which is a millionth of a Henry, or milliHenrys, mH, which is a thousandth of a Henry. Occasionally you might even find very small inductances measured in nanoHenrys, which are a thousandth of a uH.

Different Types of Inductors

Now the μ in the above equation has some interesting implications. It suggests that the magnetic field inside the inductor can be manipulated. Like mentioned above, sometimes the magnetic field created even by a solenoid falls short of requirements sometimes. That is why in almost all cases you find inductors formed around a core material.

Cores are materials that support the creation of a magnetic field. They are generally made of iron and its compounds, such as ferrite (which is an oxide of iron). You can get a greater magnetic field using a core than without one.

Air Core Inductors

Like the name suggests, this kind of inductor has no core – the core material is air! Since air has a relatively low permeability, the inductance of air core inductors is quite low – rarely above 5uH. Since they have a low inductance, the rate of current rise is quite fast for an applied voltage and that makes them capable of handling high frequencies. They are mostly used in RF circuits.

Iron Core Inductors

Iron is the most recognizable magnetic material, which makes it an ideal choice for inductors. These take the form of iron-core inductors. They are usually used for low frequency line filtering, since they can be rather beefy and have large inductances. They are also used in audio equipment.

Ferrite Core Inductors

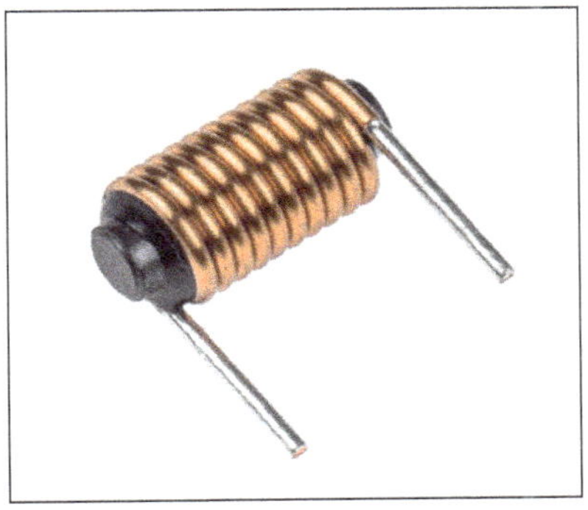

Ferrite is just a powder of oxides of iron. This powder is mixed with an epoxy resin and molded to form cores around which wires can be wound. Ferrite core inductors are easily the most recognizable because of their dull grey-black colour. They also are very brittle and break easily. They are the most widely used kinds of inductors, since the permeability can be finely controlled by controlling the ratio of ferrite to epoxy in the mix.

Inductors in Series and Parallel

Inductors connected in series and parallel behave the exact opposite way to capacitors.

For example, to calculate the inductance of a group of inductors in series, you can simply sum up the values of the individual inductances.

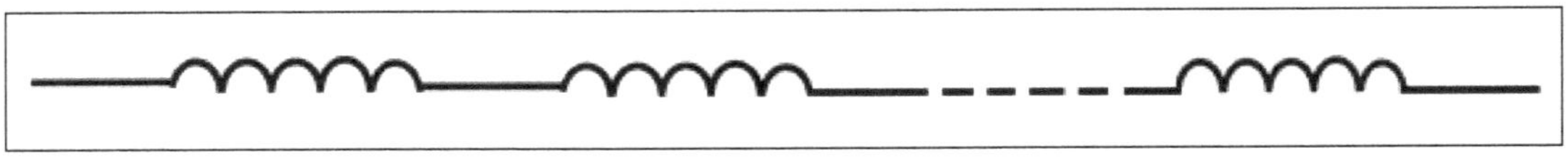

$L = L_1 + L_2 + \dots + L_n$

Where L is the total inductance and L_1, $L_2 \dots L_n$ are the individual inductances.

Suppose you have two inductors, one measuring 10uH and the other 15uH, then by putting them in series you obtain a total inductance of 25uH.

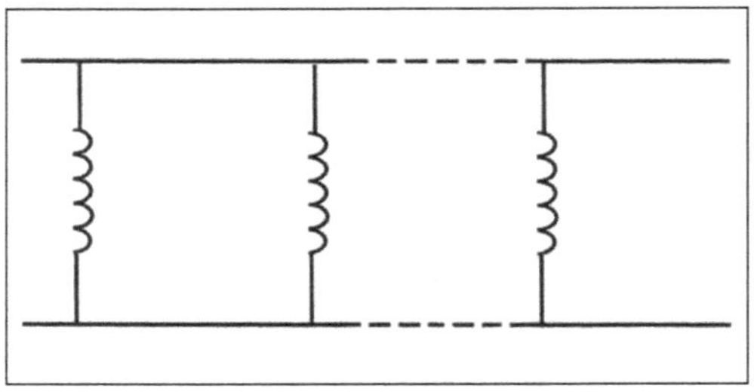

Inductors in parallel behave the same way as resistors in parallel, the inductance is given by:

$1/L = 1/L_1 + 1/L_2 + \dots\dots + 1/L_n$

Where L is the total inductance and L_1, $L_2 \dots\dots L_n$ are the individual inductances.

In this way if you connect two 10uH inductors in parallel, you'll end up with an inductance of 5uH.

Useful Inductor Formulas

Energy Stored by Inductors

Inductors can store energy much like capacitors, but the energy is gone the moment you disconnect the power and the magnetic field collapses. In other words, an unpowered inductor cannot sustain its magnetic field.

$E = ½ * L * I_2.$

Where E is the energy in Joules, L is the inductance in Henries and I is the current in amps.

If you have an inductor of 20uH with a 5A current flowing through it, then the energy stored will be 0.00025J. In this aspect inductors too, like capacitors, hold very little energy.

Rate of Current Rise

This formula has already been discussed, but it is worth having a closer look.

$V/L = dI/dt$

Where V is the voltage applied across the inductor, L is the inductance, I is the current and t is the time.

This states that when a constant voltage is applied across the inductor, the current rises in a linear slope. This can be useful in creating current ramps, just like the capacitor created voltage ramps at a constant current.

Impedance

Inductors have impedance that is related to the frequency by the formula:

$XL = 2\pi * f * L$

Where XL is the inductive impedance, f is the frequency in Hertz and L is the inductance in Henrys.

Inductor Behavior in Circuits

Surprisingly, inductors are pretty much useless in DC circuits, since there is a constant current flowing and the inductor acts like a bit of wire.

They find most of their uses in AC circuits. Like mentioned above, they have an impedance, which makes them useful to limit current in an AC circuit, such as fluorescent lamp ballasts.

They can also be used to filter signals.

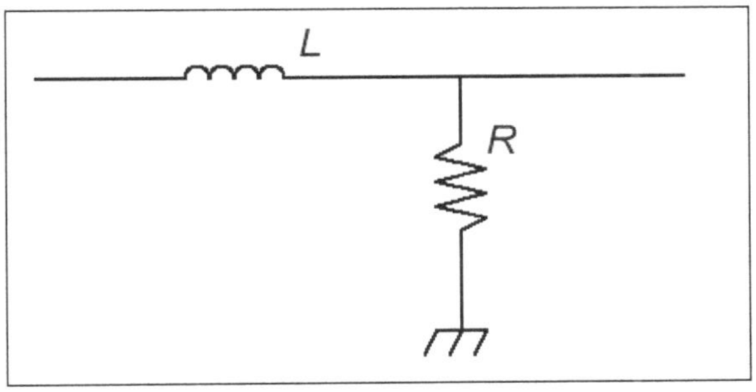

In the first case, the inductor allows all DC current to flow through it to ground, preventing all the low frequencies from reaching the output. At higher frequencies, the impedance of the inductor increases steadily, so the signal can pass through to the output, hence it is called a high pass filter.

In the second case, the inductor allows DC and low frequencies to pass through but blocks all high frequencies from the output and so it is called a low pass filter.

Inductors in Real Life

Inductors, since they are made of copper wire and ferrite, tend to be expensive and find most of their uses in radios, power supplies and telecoms equipment.

In power supplies, the property of an inductor to prevent sudden changes in current is used. Along with a capacitor, it acts to prevent sudden changes in the power supply output voltage and current.

RF circuits make use of an interesting LC circuit called a tank. The capacitor is charged and discharged into the inductor, which builds up its magnetic field. When the magnetic field collapses, a voltage is created and charges the capacitor. This creates periodic oscillations which can be used to generate high frequencies.

The frequency can be calculated by the formula:

$$f = 2\pi\sqrt{LC}$$

Where f is the frequency in Hertz, L is the inductance in Henrys and C is the capacitance in Farads.

Transistor

Transistor is a semiconductor device that can both conduct and insulate. A transistor can act as a switch and an amplifier. It converts audio waves into electronic waves and resistor, controlling electronic current. Transistors have very long life, smaller in size, can operate on lower voltage supplies for greater safety and required no filament current. The first transistor was fabricated with germanium. A transistor performs the same function as a vacuum tube triode, but using semiconductor junctions instead of heated electrodes in a vacuum chamber. It is the fundamental building block of modern electronic devices and found everywhere in modern electronic systems.

A transistor is a three terminal device. Namely,

- Base: This is responsible for activating the transistor.
- Collector: This is the positive lead.
- Emitter: This is the negative lead.

Types of Transistors

There are two types of transistors in present; they are bipolar junction transistor (BJT), field effect transistors (FET). A small current is flowing between the base and the emitter; base terminal can control a larger current flow between the collector and the emitter terminals. For a field-effect transistor, it also has the three terminals, they are gate, source, and drain, and a voltage at the gate can control a current between source and drain.

The simple diagrams of BJT and FET are shown in figure below:

Bipolar Junction Transistor(BJT)

Field Effect Transistors(FET)

Transistors come in a variety of different sizes and shapes. One thing all of these transistors have in common is that they each have three leads.

Bipolar Junction Transistor

A Bipolar Junction Transistor (BJT) has three terminals connected to three doped semiconductor regions. It comes with two types, P-N-P and N-P-N.

P-N-P transistor, consisting of a layer of N-doped semiconductor between two layers of P-doped material. The base current entering in the collector is amplified at its output.

That is when PNP transistor is ON when its base is pulled low relative to the emitter. The arrows of PNP transistor symbol the direction of current flow when the device is in forward active mode.

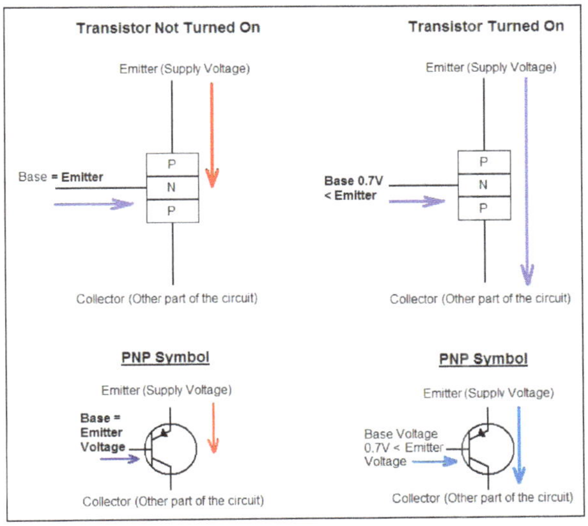

N-P-N transistor consisting a layer of P-doped semiconductor between two layers of N-doped material. By amplifying current the base we get the high collector and emitter current.

That is when NPN transistor is ON when its base is pulled low relative to the emitter. When the transistor is in ON state, current flow is in between the collector and emitter of the transistor. Based on minority carriers in P-type region the electrons moving from emitter to collector. It allows the greater current and faster operation; because of this reason most bipolar transistors used today are NPN.

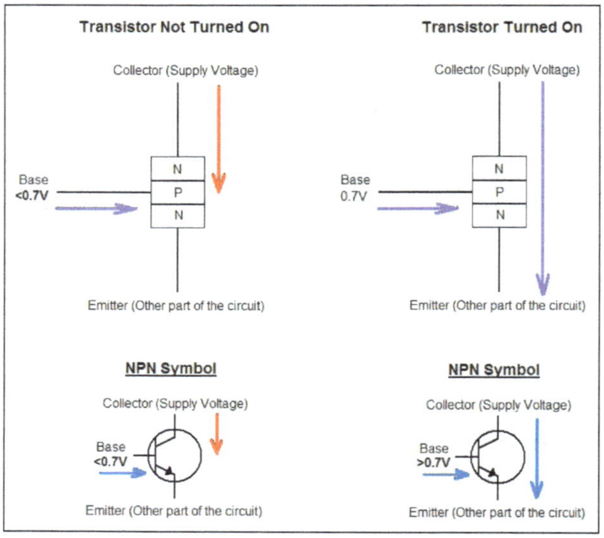

Field Effect Transistor (FET)

The field effect transistor is a unipolar transistor, N-channel FET or P-channel FET are used for conduction. The three terminals of FET are source, gate and drain. The basic n-channel and p-channel FET's are shown above. For an n-channel FET, the device is constructed from n-type material. Between the source and drain then-type material acts as a resistor.

This transistor controls the positive and negative carriers with respect to holes or electrons. FET channel is formed by moving of positive and negative charge carriers. The channel of FET which is made by silicon.

There are many types of FET's, MOSFET, JFET and etc. The applications of FET's are in low noise amplifier, buffer amplifier and analog switch.

Bipolar Junction Transistor Biasing

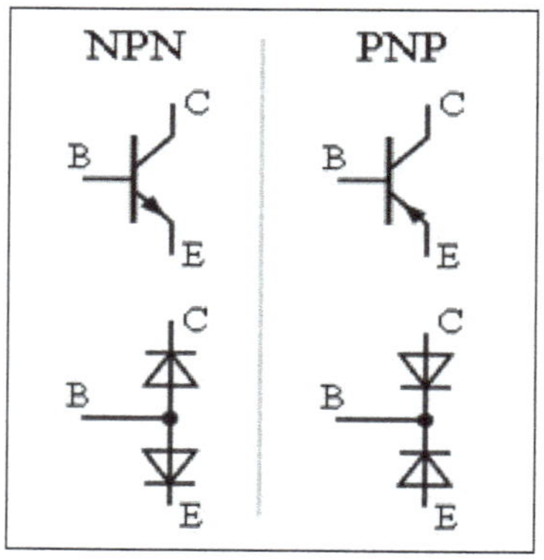

Transistors are the most important semiconductor active devices essential for almost all circuits. They are used as electronic switches, amplifiers etc. in circuits. Transistors may be NPN, PNP, FET, JFET etc. which have different functions in electronic circuits. For the proper working of the circuit, it is necessary to bias the transistor using resistor networks. Operating point is the point on the output characteristics that shows the Collector-Emitter voltage and the Collector current with no input signal. The Operating point is also known as the Bias point or Q-Point (Quiescent point).

Biasing is referred to provide resistors, capacitors or supply voltage etc. to provide proper operating characteristics of the transistors. DC biasing is used to obtain DC collector current at a particular collector voltage. The value of this voltage and current are expressed in terms of the Q-Point. In a transistor amplifier configuration, the IC (max) is the maximum current that can flow through the transistor and VCE (max) is the maximum voltage applied across the device. To work the transistor as an amplifier, a load resistor RC must be connected to the collector. Biasing set the DC operating voltage and current to the correct level so that the AC input signal can be properly amplified by the transistor. The correct biasing point is somewhere between the fully ON or fully OFF states of the transistor. This central point is the Q-Point and if the transistor is properly biased, the Q-point will be the central operating point of the transistor. This helps the output current to increase and decrease as the input signal swings through the complete cycle.

For setting the correct Q-Point of the transistor, a collector resistor is used to set the collector current to a constant and steady value without any signal in its base. This steady DC operating point is set by the value of the supply voltage and the value of the base biasing resistor. Base bias resistors are used in all the three transistor configurations like common base, common collector and Common emitter configurations.

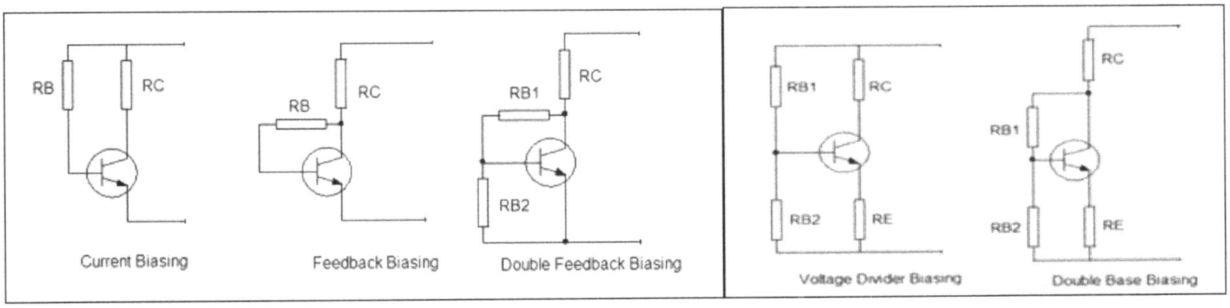

Modes of Biasing

Following are the different modes of transistor base:

Current Biasing

As shown in the figure of current biasing, two resistors RC and RB are used to set the base bias. These resistors establish the initial operate in region of the transistor with a fixed current bias.

The transistor forward biases with a positive base bias voltage through RB. The forward base-Emitter voltage drop is 0.7 volts. Therefore the current through RB is $I_B = (V_{cc} - V_{BE}) / I_B$.

Feedback Biasing

Figure(Feedback biasing)shows the transistor biasing by the use of a feedback resistor. The base bias is obtained from the collector voltage. The collector feedback ensures that the transistor is always biased in the active region. When the collector current increases, the voltage at the collector drops. This reduces the base drive which in turn reduces the collector current. This feedback configuration is ideal for transistor amplifier designs.

Double Feedback Biasing

Figure(Double Feedback Biasing) shows how the biasing is achieved using double feedback resistors.

Voltage Dividing Biasing

Figure(Voltage divider biasing) shows the Voltage divider biasing in which two resistors RB1 and RB2 are connected to the base of the transistor forming a voltage divider network. The transistor gets biases by the voltage drop across RB2. This kind of biasing configuration is used widely in amplifier circuits.

Double Base Biasing

Figure(double feedback biasing) shows a double feedback for stabilization. It uses both Emitter and Collector base feedback to improve the stabilization through controlling the collector current. Resistor values should be selected so as to set the voltage drop across the Emitter resistor 10% of the supply voltage and the current through RB1, 10% of the collector current.

Advantages of Transistor

1. Smaller mechanical sensitivity.
2. Lower cost and smaller in size, especially in small-signal circuits.
3. Low operating voltages for greater safety, lower costs and tighter clearances.
4. Extremely long life.
5. No power consumption by a cathode heater.
6. Fast switching.

Diode

A diode is defined as a two-terminal electronic component that only conducts current in one direction (so long as it is operated within a specified voltage level). An ideal diode will have zero resistance in one direction, and infinite resistance in the reverse direction.

Although in the real world, diode can't achieve zero or infinite resistance. Instead, a diode will have negligible resistance in one direction (to allow current flow), and a very high resistance in the reverse direction (to prevent current flow). A diode is effectively like a valve for an electrical circuit.

Semiconductor diodes are the most common type of diode. These diodes begin conducting electricity only if a certain threshold voltage is present in the forward direction (i.e. the "low resistance" direction). The diode is said to be "forward biased" when conducting current in this direction. When connected within a circuit in the reverse direction (i.e. the "high resistance" direction), the diode is said to be "reverse biased".

A diode only blocks current in the reverse direction (i.e. when it is reverse biased) while the reverse voltage is within a specified range. Above this range, the reverse barrier breaks. The voltage at which this breakdown occurs is called the "reverse breakdown voltage". When the voltage of the circuit is higher than the reverse breakdown voltage, the diode is able to conduct electricity in the reverse direction (i.e. the "high resistance" direction). This is why in practice we say diode's have a high resistance in the reverse direction – not an infinite resistance.

A PN junction is the simplest form of the semiconductor diode. In ideal conditions, this PN junction behaves as a short circuit when it is forward biased and as an open circuit when it is in the reverse biased. The name diode is derived from "diode" which means a device that has two electrodes.

Diode Symbol

The symbol of a diode is shown below. The arrowhead points in the direction of conventional current flow in the forward biased condition. That means the anode is connected to the p side and cathode is connected to the n side.

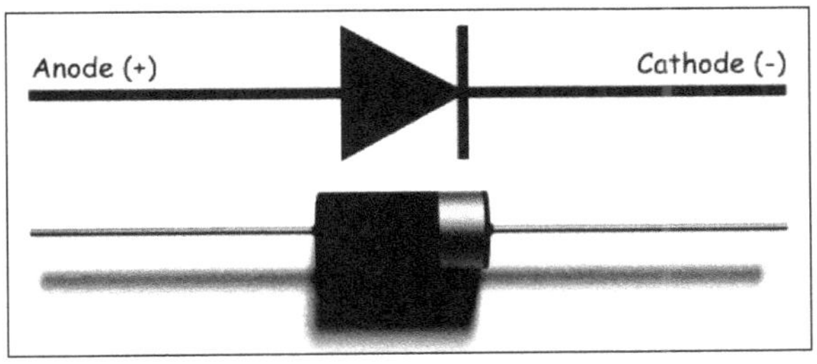

We can create a simple PN junction diode by doping pentavalent or donor impurity in one portion and trivalent or acceptor impurity in other portion of silicon or germanium crystal block. These dopings make a PN junction at the middle part of the block. We can also form a PN junction by joining a p-type and n-type semiconductor together with a special fabrication technique. The terminal connected to the p-type is the anode. The terminal connected to the n-type side is the cathode.

Working Principle of Diode

A diode's working principle depends on the interaction of n-type and p-type semiconductors. An n-type semiconductor has plenty of free electrons and a very few numbers of holes. In other words, we can say that the concentration of free electrons is high and that of holes is very low in an n-type semiconductor. Free electrons in the n-type semiconductor are referred as majority charge carriers, and holes in the n-type semiconductor are referred to as minority charge carriers.

A p-type semiconductor has a high concentration of holes and low concentration of free electrons. Holes in the p-type semiconductor are majority charge carriers, and free electrons in the p-type semiconductor are minority charge carriers.

Unbiased Diode

When one n-type region and one p-type region come in contact, then due to concentration difference, majority carriers diffuse from one side to another. As the concentration of holes is high in the p-type region and it is low in the n-type region, the holes start diffusing from the p-type region to n-type region. Again the concentration of free electrons is high in the n-type region and it is low in the p-type region and due to this reason, free electrons start diffusing from the n-type region to the p-type region.

The free electrons diffusing into the p-type region from the n-type region would recombine with holes available there and create uncovered negative ions in the p-type region. In the same way, the holes diffusing into the n-type region from the p-type region would recombine with free electrons available there and create uncovered positive ions in the n-type region.

In this way, there would a layer of negative ions in the p-type side and a layer of positive ions in the n-type region appear along the junction line of these two types of semiconductor. The layers of uncovered positive ions and uncovered negative ions form a region at the middle of the diode where no charge carrier exists since all the charge carriers get recombined here in this region. Due to lack of charge carriers, this region is called depletion region.

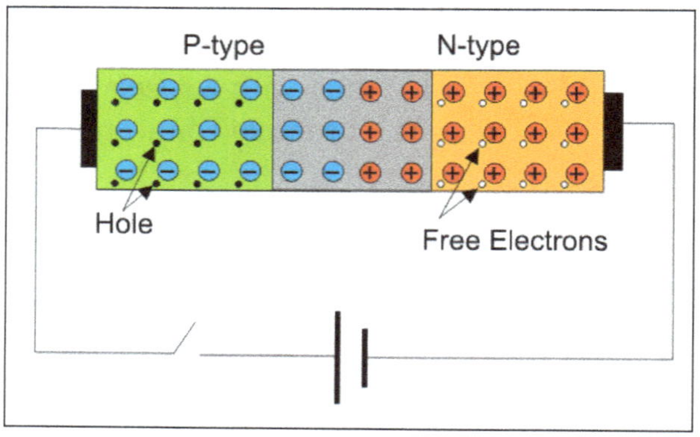

After the formation of the depletion region, there is no more diffusion of charge carriers from one side to another in the diode. This is because due to the electric field appeared across the depletion region will prevent further migration of charge carriers from one side to another. The potential of the layer of uncovered positive ions in the n-type side would repeal the holes in the p-type side and the potential of the layer of uncovered negative ions in the p-type side would repeal the free electrons in the n-type side. That means a potential barrier is created across the junction to prevent further diffusion of charge carriers.

Forward Biased Diode

In forward biased diode, there is no current flowing through the diode. This is because although there is an external electrical field applied across the diode but still the majority charge carriers do not get sufficient influence of the external field to cross the depletion region. As we told that the depletion region acts as a potential barrier against the majority charge carriers. This potential barrier is called forward potential barrier. The majority charge carriers start crossing the forward potential barrier only when the value of externally applied voltage across the junction is more than the potential of the forward barrier. For silicon diodes, the forward barrier potential is 0.7 volt and for germanium diodes, it is 0.3 volt. When the externally applied forward voltage across the diode becomes more than the forward barrier potential, the free majority charge carriers start crossing the barrier and contribute the forward diode current. In that situation, the diode would behave as a short-circuited path and the forward current gets limited by only externally connected resistors to the diode.

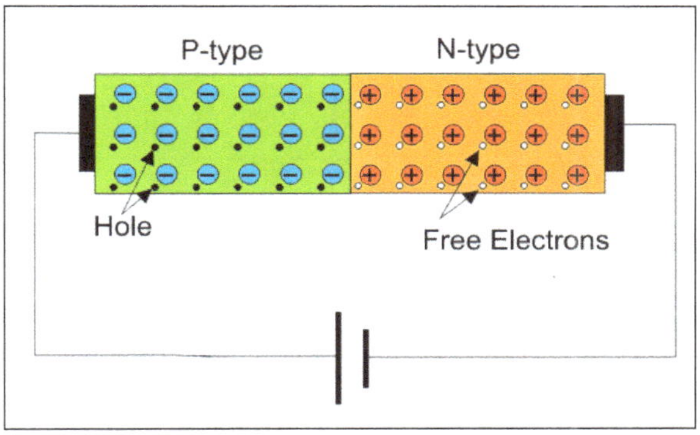

Reverse Biased Diode

In Reverse Biased condition, the negative terminal of the voltage source is connected to the p-type side and positive terminal of the voltage is connected source to the n-type side of the diode. At that condition, due to electrostatic attraction of negative potential of the source, the holes in the p-type region would be shifted more away from the junction leaving more uncovered negative ions at the junction. In the same way, the free electrons in the n-type region would be shifted more away from the junction towards the positive terminal of the voltage source leaving more uncovered positive ions in the junction. As a result of this phenomenon, the depletion region becomes wider. This condition of a diode is called the reverse biased condition. At that condition, no majority carriers cross the junction as they go away from the junction. In this way, a diode blocks the flow of current when it is reverse biased.

There are always some free electrons in the p-type semiconductor and some holes in the n-type semiconductor. These opposite charge carriers in a semiconductor are called minority charge carriers. In the reverse biased condition, the holes find themselves in the n-type side would easily cross the reverse biased depletion region as the field across the depletion region does not present rather it helps minority charge carriers to cross the depletion region. As a result, there is a tiny current flowing through the diode from positive to the negative side. The amplitude of this current is very small as the number of minority charge carriers in the diode is very small. This current is called reverse saturation current.

If the reverse voltage across a diode gets increased beyond a safe value, due to higher electrostatic force and due to a higher kinetic energy of minority charge carriers colliding with atoms, a number of covalent bonds get broken to contribute a huge number of free electron-hole pairs in the diode and the process is cumulative. The huge number of such generated charge carriers would contribute a huge reverse current in the diode. If this current is not limited by an external resistance connected to the diode circuit, the diode may permanently be destroyed.

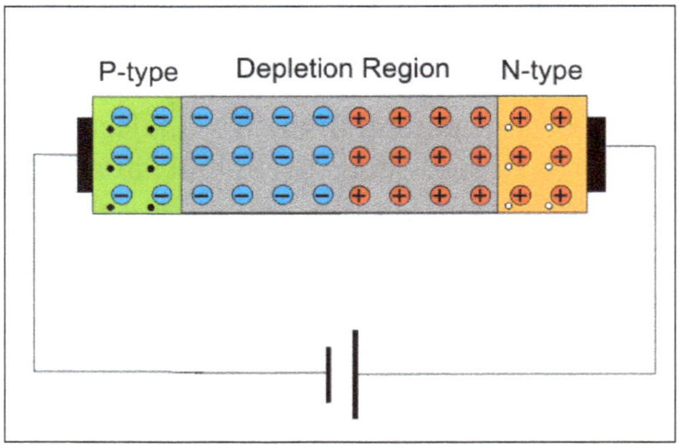

Types of Diodes

The other variant of diodes have different construction, characteristics and applications. The different types of diodes are:

- Small signal or small current diode - These diodes assumes that the operating point is not affected because the signal is small.
- Large signal diodes - The operating point in these diodes get affected as the signal is large.
- Zener diodes - This diode runs in reverse bias condition when the voltage reaches the breakdown point. A stable voltage can be achieved by placing a resistor across it to limit the current. This diode is used to provide reference voltage in power supply circuits.
- Light emitting diodes (LED) - This is the most popular kind of diode. When it works in the forward bias condition, the current flows through the junction to produce the light.
- Photodiodes - The electrons and holes are generated as light strikes across the p-n junction causing the current to flow. Theses diodes can work as photo detector and are used to generate electricity.
- Constant current diodes - This diode keeps the current constant even when the voltage applied keeps changing. It consists of JFET (junction – field effect transistor) with the source shorted to the gate in order to function like a two - terminal current limiter or current source.

- Schottky diode - These diodes are used in RF applications and clamping circuits. This diode has lower forward voltage drop as against the silicon PN junction diodes.
- Shockley diode - This is a four layer diode which is also known as PNPN diode. This didoe is similar to thermistor where the gate is disconnected.
- Step recovery diodes - This semiconductor diode has the ability to generate short pulses and hence it is used in microwave applications as a pulse generator.
- Tunnel diodes - This diode is heavily doped in the forward bias condition that has a negative resistance at extremely low voltage and a short circuit in the negative bias direction. This diode is useful as a microwave amplifier and in oscillators.
- Varactor diodes - This diode works in reverse bias condition and restricts, the flow of current through the junction. Depending on the amount of biasing, the width of the depletion region keeps varying. This diode comprises of two plates of a capacitor with the depletion region amidst them. The variation in capacitance depends upon the depletion region and this can varied by altering the reverse bias on the diode.
- PIN diodes - This diode has intrinsic semiconductor sandwiched between P- type and N-type region. Doping does not occur in this type of diode and thereby the intrinsic semiconductor increases the width of the depletion region. They are used as photodiodes and radio frequency switches.
- LASER diode - This diode produces laser type of light and are expensive as compared to LED. They are widely used in CD and DVD drives.
- Transient voltage suppression diodes - This diode is used to protect the electronics that are sensitive against voltage spikes.
- Gold doped diodes - These diodes use gold as the dopant and can operate at signal frequencies even if the forward voltage drop increases.
- Super barrier diodes - These are also called as the rectifier diodes. This diodes have the property of low reverse leakage current as that of normal p-n junction diode and low forward voltage drop as that of Schottky diode with surge handling ability.
- Point contact diodes - The construction of this diode is simpler and are used in analogy applications and as a detector in radio receivers. This diode is built of n – type semiconductor and few conducting metals placed to be in contact with the semiconductor. Some metals move from towards the semiconductor to form small region of p- type semiconductor near the contact.
- Pettier diodes - This diode is used as heat engine and sensor for thermoelectric cooling.
- Gunn diode - This diode is made of materials like Gas or INP that exhibit a negative differential resistance region.
- Crystal diode - These are a type of point contact diodes which are also called as Cat's whisker diode. This diode comprises of a thin sharpened metal wire which is pressed against the semiconducting crystal. The metal wire is the anode and the semiconducting crystal is the cathode. These diodes are obsolete.

- Avalanche diode - This diode conducts in reverse bias condition where the reverse bias voltage applied across the p-n junction creates a wave of ionization leading to the flow of large current. These didoes are designed to breakdown at specific reverse voltage in order to avoid any damage.
- Silicon controlled rectifier - As the name implies this diode can be controlled or triggered to the ON condition due to the application of small voltage. They belong to the family of Thermistors and is used in various fields of DC motor control, generator field regulation, lighting system control and variable frequency drive . This is three terminal device with anode, cathode and third controlled lead or gate.
- Vacuum diodes - This diode is two electrode vacuum tube which can tolerate high inverse voltages.

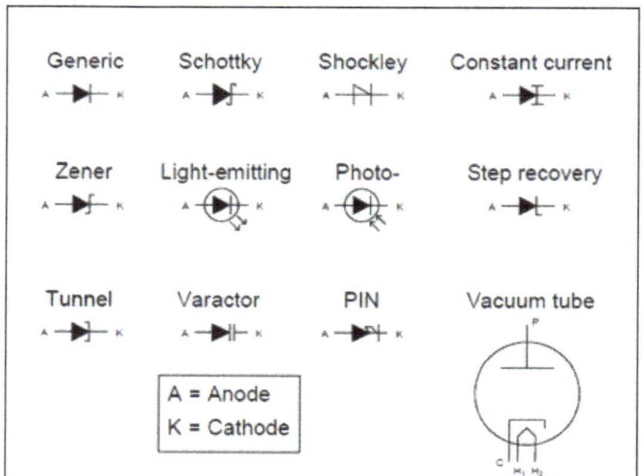

Various symbols of Diode

Image showing symbols of various types of diodes

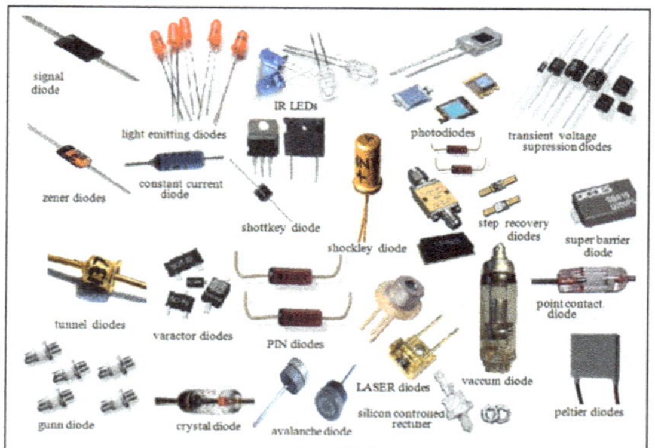

Types of Diode

Image showing various types of diodes

Generic Diodes

Small Signal and Large Signal

A p-n junction diode is the simplest semiconductor device. It is a two-terminal, bipolar, unilateral rectifying device that conducts only in one direction. The generic diodes are used in the following fields:

- Rectification in power supply circuits.

- Extraction of modulation from radio signals in a radio receiver and in protection circuits where large transient currents may appear on low current transistors or ICs in interfacing with relays or other high power devices.
- Used in series with power inputs to electronic circuits where only one of negative or positive polarity voltage is desired.

Voltage Source

A Voltage Source is a device that generates an exact output voltage which, in theory, does not change regardless of the load current.

A voltage source, such as a battery or generator, provides a potential difference (voltage) between two points within an electrical circuit allowing current to flowing around it. Remember that voltage can exist without current. A battery is the most common voltage source for a circuit with the voltage that appears across the positive and negative terminals of the source being called the terminal voltage.

Ideal Voltage Source

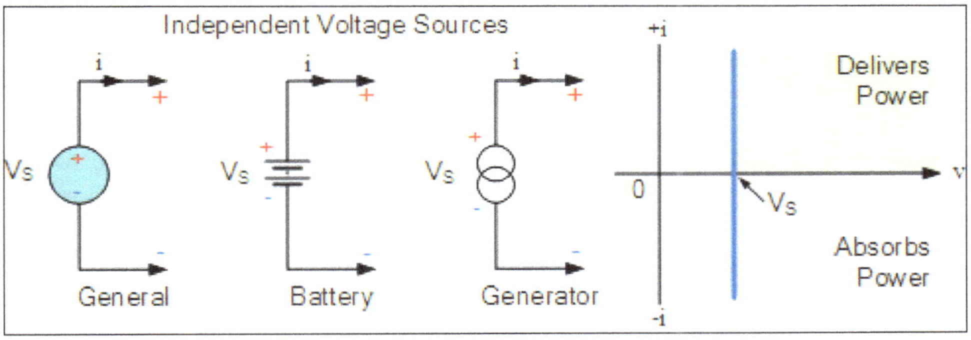

An ideal voltage source is defined as a two terminal active element that is capable of supplying and maintaining the same voltage, (v) across its terminals regardless of the current, (i) flowing through it. In other words, an ideal voltage source will supply a constant voltage at all times regardless of the value of the current being supplied producing an I-V characteristic represented by a straight line.

Then an ideal voltage source is known as an Independent Voltage Source as its voltage does not depend on either the value of the current flowing through the source or its direction but is determined solely by the value of the source alone. So for example, an automobile battery has a 12V terminal voltage that remains constant as long as the current through it does not become to high, delivering power to the car in one direction and absorbing power in the other direction as it charges.

On the other hand, a Dependent Voltage Source or controlled voltage source, provides a voltage supply whose magnitude depends on either the voltage across or current flowing through some other circuit element. A dependent voltage source is indicated with a diamond shape and are used as equivalent electrical sources for many electronic devices, such as transistors and operational amplifiers.

Connecting Voltage Sources Together

Ideal voltage sources can be connected together in both parallel or series the same as for any circuit element. Series voltages add together while parallel voltages have the same value. Unequal ideal voltage sources cannot be connected directly together in parallel.

Voltage Source in Parallel

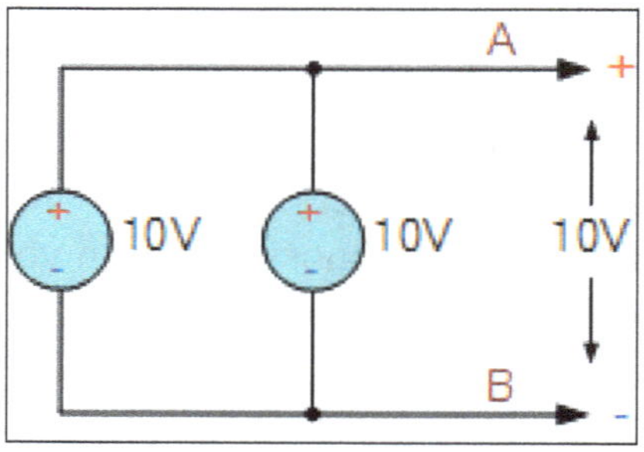

Ideal voltage sources can be connected in parallel provided they are of the same voltage value. Here in this example, two 10 volt voltage source are combined to produce 10 volts between terminals A and B. Ideally, there would be just one single voltage source of 10 volts given between terminals A and B.

What is not allowed or is not best practice, is connecting together ideal voltage sources that have different voltage values as shown, or are short-circuited by an external closed loop or branch.

Badly Connected Voltage Sources

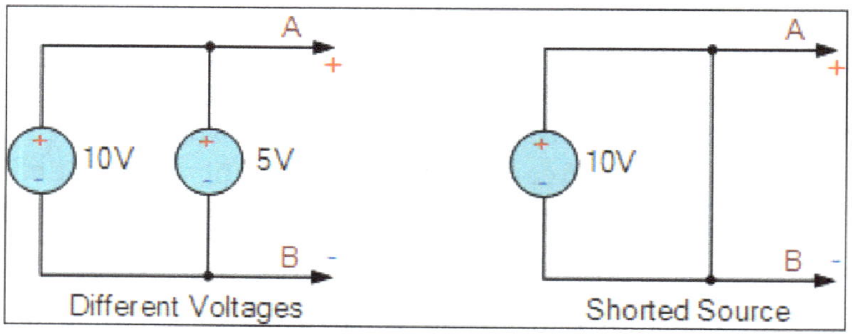

However, when dealing with circuit analysis, voltage sources of different values can be used providing there are other circuit elements in between them to comply with Kirchoff's Voltage Law, KVL.

Unlike parallel connected voltage sources, ideal voltage sources of different values can be connected together in series to form a single voltage source whose output will be the algebraic addition or subtraction of the voltages used. Their connection can be as: series-aiding or series-opposing voltages as shown.

Voltage Source in Series

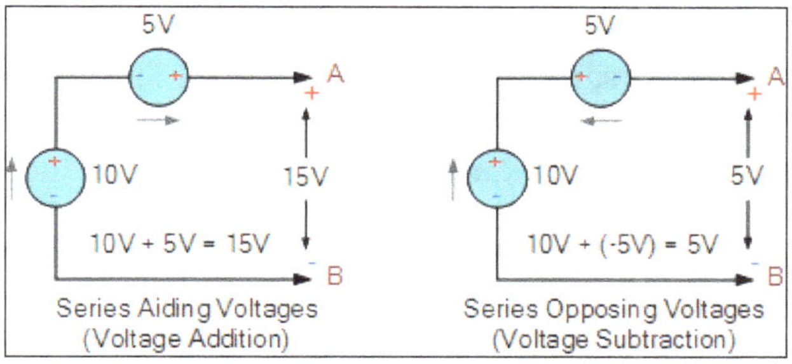

Series aiding voltage sources are series connected sources with their polarities connected so that the plus terminal of one is connected to the negative terminal of the next allowing current to flow in the same direction. In the example above, the two voltages of 10V and 5V of the first circuit can be added, for a V_S of 10 + 5 = 15V. So the voltage across terminals A and B is 15 volts.

Series opposing voltage sources are series connected sources which have their polarities connected so that the plus terminal or the negative terminals are connected together as shown in the second circuit above. The net result is that the voltages are subtracted from each other. Then the two voltages of 10V and 5V of the second circuit are subtracted with the smaller voltage subtracted from the larger voltage. Resulting in a V_S of 10 - 5 = 5V.

The polarity across terminals A and B is determined by the larger polarity of the voltage sources, in this example terminal A is positive and terminal B is negative resulting in +5 volts. If the series-opposing voltages are equal, the net voltage across A and B will be zero as one voltage balances out the other. Also any currents (I) will also be zero, as without any voltage source, current can not flow.

Voltage Source Examples

1. Two series aiding ideal voltage sources of 6 volts and 9 volts respectively are connected together to supply a load resistance of 100 Ohms. Calculate: the source voltage, V_S, the load current through the resistor, I_R and the total power, P dissipated by the resistor. Draw the circuit.

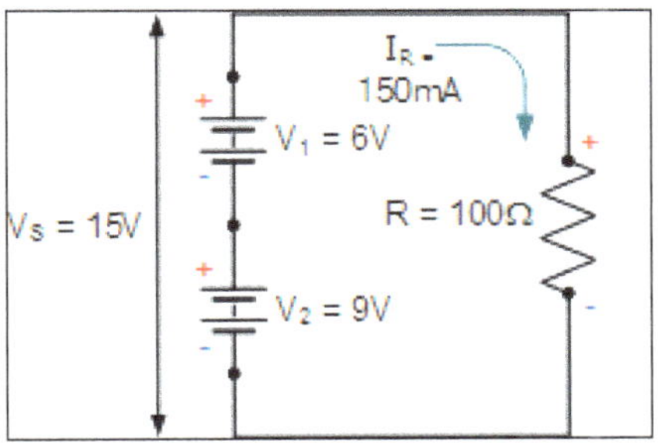

$$V_S = V_1 + V_2 = 6+9 = 15V$$

$$I_R = \frac{V_S}{R} = \frac{15V}{100\Omega} = 150mA$$

$$P_R = I^2R = 0.15^2 \times 100\, 2.25W$$

Thus, V_S = 15V, I_R = 150mA or 0.15A, and P_R = 2.25W.

Practical Voltage Source

An ideal voltage source can provide a voltage supply that is independent of the current flowing through it, that is, it maintains the same voltage value always. This idea may work well for circuit analysis techniques, but in the real world voltage sources behave a little differently as for a practical voltage source, its terminal voltage will actually decrease with an increase in load current.

As the terminal voltage of an ideal voltage source does not vary with increases in the load current, this implies that an ideal voltage source has zero internal resistance, $R_S = 0$. In other words, it is a resistor less voltage source. In reality all voltage sources have a very small internal resistance which reduces their terminal voltage as they supply higher load currents.

For non-ideal or practical voltage sources such as batteries, their internal resistance (R_S) produces the same effect as a resistance connected in series with an ideal voltage source as these two series connected elements carry the same current as shown.

Ideal and Practical Voltage Source

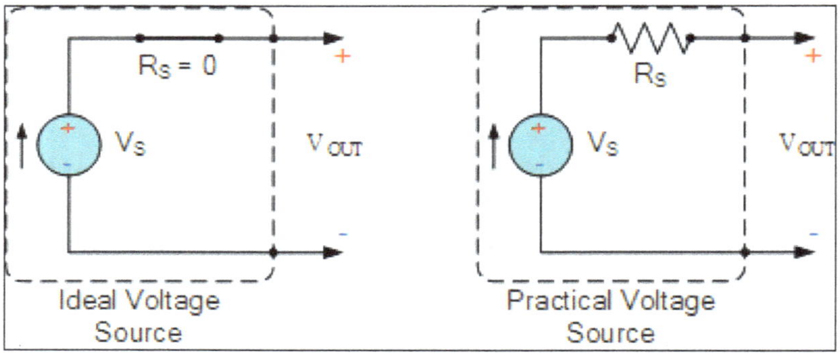

A practical voltage source closely resembles that of a Thevenin's equivalent circuit as Thevenin's theorem states that "any linear network containing resistances and sources of emf and current may be replaced by a single voltage source, V_S in series with a single resistance, R_S "if the series source resistance is low, the voltage source is ideal. When the source resistance is infinite, the voltage source is open-circuited.

In the case of all real or practical voltage sources, this internal resistance, R_S no matter how small has an effect on the I-V characteristic of the source as the terminal voltage falls off with an increase in load current. This is because the same load current flows through R_S.

Ohms law tells us that when a current flows through a resistance, a voltage drop is produce across the same resistance. The value of this voltage drop is given as $i*R_S$. Then V_{OUT} will equal the ideal voltage source, V_S minus the $i*R_S$ voltage drop across the resistor. Remember that in the case of an ideal source voltage, R_S is equal to zero as there is no internal resistance, therefore the terminal voltage is same as V_S.

Then the voltage sum around the loop given by Kirchoff's voltage law, KVL is: $V_{OUT} = V_S - i*R_S$. This equation can be plotted to give the I-V characteristics of the actual output voltage. It will give a straight line with a slope $-R_S$ which intersects the vertical voltage axis at the same point as V_S when the current $i = 0$ as shown.

Practical Voltage Source Characteristics

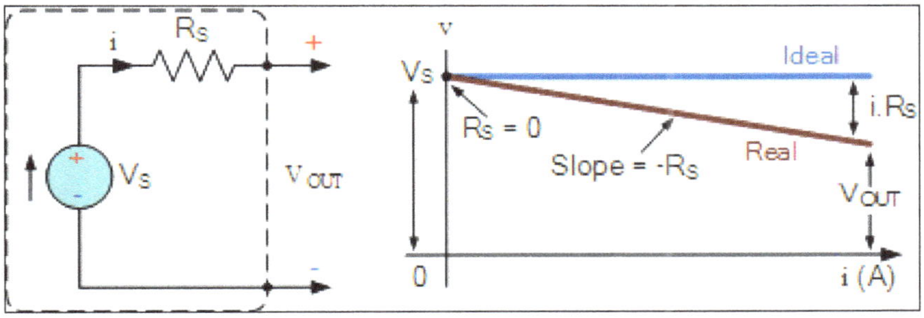

Therefore, all ideal voltage sources will have a straight line I-V characteristic but non-ideal or real practical voltage sources will not but instead will have an I-V characteristic that is slightly angled down by an amount equal to $i*R_S$ where R_S is the internal source resistance (or impedance). The I-V characteristics of a real battery provides a very close approximation of an ideal voltage source since the source resistance R_S is usually quite small.

1. The decrease in the angle of the slope of the I-V characteristics as the current increases is known as regulation. Voltage regulation is an important measure of the quality of a practical voltage source as it measures the variation in terminal voltage between no load, that is when $I_L = 0$, (an open-circuit) and full load, that is when I_L is at maximum, (a short-circuit).

2. A battery supply consists of an ideal voltage source in series with an internal resistor. The voltage and current measured at the terminals of the battery were found to be $V_{OUT1} = 130V$ at 10A, and $V_{OUT2} = 100V$ at 25A. Calculate the voltage rating of the ideal voltage source and the value of its internal resistance. Draw the I-V characteristics.

Firstly lets define in simple "simultaneous equation form", the two voltage and current outputs of the battery supply given as: V_{OUT1} and V_{OUT2}.

$$V_{OUT} = V_s - iR_s$$

$$V_{OUT1} = 130 = V_s + 10R_s$$

$$V_{OUT2} = 100 = V_s + 25R_s$$

As with have the voltages and currents in a simultaneous equation form, to find V_S we will first multiply V_{OUT1} by five, (5) and V_{OUT2} by two, (2) as shown to make the value of the two currents, (i) the same for both equations.

$$V_{OUT1} = 130 = V_s + 10Rs \text{ ....} \times 5$$
$$V_{OUT2} = 100 = V_s + 25Rs \text{ ....} \times 2$$

$$V_{OUT1} = 650 = 5V_s + 50R_s$$
$$V_{OUT2} = 200 = 2V_s + 50R_s$$

Having made the co-efficients for R_S the same by multiplying through with the previous constants, we now multiply the second equation V_{OUT2} by minus one, (-1) to allow for the subtraction of the two equations so that we can solve for V_S as shown.

$$V_{OUT1} = 650 = 5V_s + 50R_s$$
$$V_{OUT2} = 200 = 2V_s + 50R_s .... \times -1$$

$$V_{OUT1} = 650 = 5V_S + 50R_S$$
$$V_{OUT2} = -200 = -2V_S - 50R_S$$

Rearrange to give:

$$650 - 200 = (5V_S - 2Vs) + (50R_S - 50R_S)$$
$$450 = 3\,Vs + 0$$
$$\therefore Vs = 45° = 150V$$

Knowing that the ideal voltage source, V_S is equal to 150 volts, we can use this value for equation V_{OUT1} (or V_{OUT2} if so wished) and solve to find the series resistance, R_S.

$$V_{OUT1} — 130 = V_s + 10R_s$$

$$V_s = 150V$$

$$130 = 150 + 10R_S$$

$$\therefore R_S = \frac{150-130}{10} = 2\Omega$$

Then for our simple example, the batteries internal voltage source is calculated as: V_S = 150 volts, and its internal resistance as: $R_S = 2\Omega$. The I-V characteristics of the battery are given as:

Battery I-V Characteristics

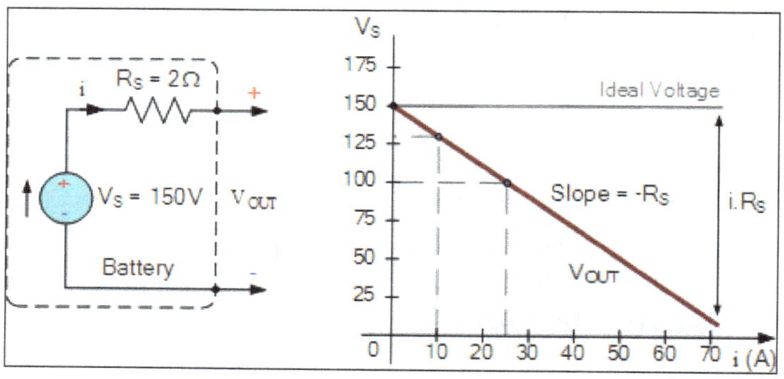

Dependent Voltage Source

Unlike an ideal voltage source which produces a constant voltage across its terminals regardless of what is connected to it, a controlled or dependent voltage source changes its terminal voltage depending upon the voltage across, or the current through, some other element connected to the circuit, and as such it is sometimes difficult to specify the value of a dependent voltage source,

Dependent voltage sources behave similar to the electrical sources we have looked at so far, both practical and ideal (independent) the difference this time is that a dependent voltage source can be controlled by an input current or voltage. A voltage source that depends on a voltage input is generally referred to as a Voltage Controlled Voltage Source or VCVS. A voltage source that depends on a current input is referred to as a Current Controlled Voltage Source or CCVS.

Ideal dependent sources are commonly used in the analysing the input/output characteristics or the gain of circuit elements such as operational amplifiers, transistors and integrated circuits. Generally, an ideal voltage dependent source, either voltage or current controlled is designated by a diamond-shaped symbol as shown.

Dependent Voltage Source Symbols

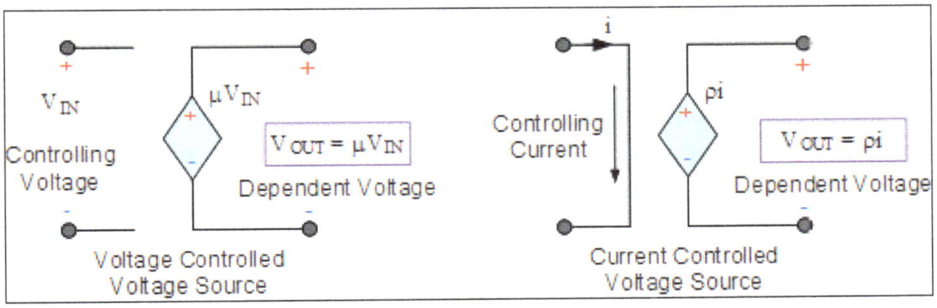

An ideal dependent voltage-controlled voltage source, VCVS, maintains an output voltage equal to some multiplying constant (basically an amplification factor) times the controlling voltage present elsewhere in the circuit. As the multiplying constant is, well, a constant, the controlling voltage, V_{IN} will determine the magnitude of the output voltage, V_{OUT}. In other words, the output voltage "depends" on the value of input voltage making it a dependent voltage source and in many ways, an ideal transformer can be thought of as a VCVS device with the amplification factor being its turns ratio.

Then the VCVS output voltage is determined by the following equation: $V_{OUT} = \mu V_{IN}$. The multiplying constant μ is dimensionless as it is purely a scaling factor because $\mu = V_{OUT}/V_{IN}$, so its units will be volts/volts.

An ideal dependent current-controlled voltage source, CCVS, maintains an output voltage equal to some multiplying constant (rho) times a controlling current input generated elsewhere within the connected circuit. Then the output voltage "depends" on the value of the input current, again making it a dependent voltage source.

As a controlling current, I_{IN} determines the magnitude of the output voltage, V_{OUT} times the magnification constant ρ (rho), this allows us to model a current-controlled voltage source as a trans-resistance amplifier as the multiplying constant, ρ gives us the following equation: $V_{OUT} = \rho I_{IN}$. This multiplying constant ρ (rho) has the units of Ohm's because $\rho = V_{OUT}/I_{IN}$, and its units will therefore be volts/amperes.

Current Source

A Current Source is an active circuit element that is capable of supplying a constant current flow to a circuit regardless of the voltage developed across its terminals.

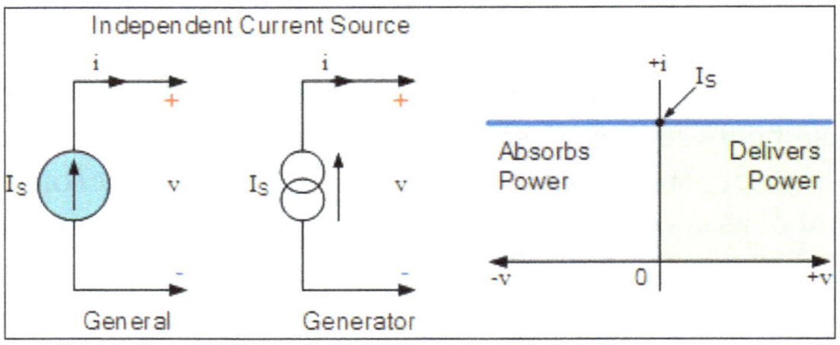

An ideal current source is called a "constant current source" as it provides a constant steady state current independent of the load connected to it producing an I-V characteristic represented by a straight line. As with voltage sources, the current source can be either independent (ideal) or dependent (controlled) by a voltage or current elsewhere in the circuit, which itself can be constant or time-varying.

Ideal independent current sources are typically used to solve circuit theorems and for circuit analysis techniques for circuits that containing real active elements. The simplest form of a current source is a resistor in series with a voltage source creating currents ranging from a few milli-amperes to many hundreds of amperes. Remember that a zero-value current source is an open circuit as R = 0.

The concept of a current source is that of a two-terminal element that allows the flow of current indicated by the direction of the arrow. Then a current source has a value, i, in units of amperes, (A) which are typically abbreviated to amps. The physical relationship between a current source and voltage variables around a network is given by Ohm's law as these voltage and current variables will have specified values.

It may be difficult to specify the magnitude and polarity of voltage of an ideal current source as a function of the current especially if there are other voltage or current sources in the connected circuit. Then we may know the current supplied by the current source but not the voltage across it unless the power supplied by the current source is given,

$P = V*I$.

However, if the current source is the only source within the circuit, then the polarity of voltage across the source will be easier to establish. If however there is more than one source, then the terminal voltage will be dependent upon the network in which the source is connected.

Connecting Current Sources Together

Just like voltage sources, ideal current sources can also be connected together to increase (or decrease) the available current. But there are rules on how two or more independent current sources with different values can be connected, either in series or parallel.

Current Source in Parallel

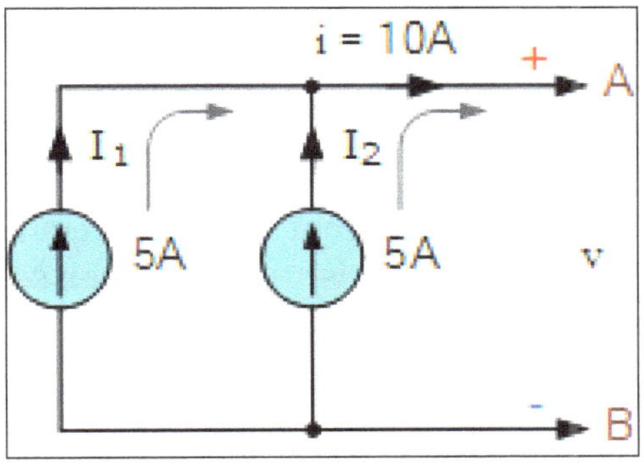

Connecting two or more current sources in parallel is equivalent to one current source whose total current output is given as the algebraic addition of the individual source currents. Here in this example, two 5 amp current sources are combined to produce 10 amps as $I_T = I_1 + I_2$.

Current sources of different values may be connected together in parallel. For example, one of 5 amps and one of 3 amps would combined to give a single current source of 8 amperes as the arrows representing the current source both point in the same direction. Then as the two currents add together, their connection is said to be: parallel-aiding.

Parallel Opposing Current Sources

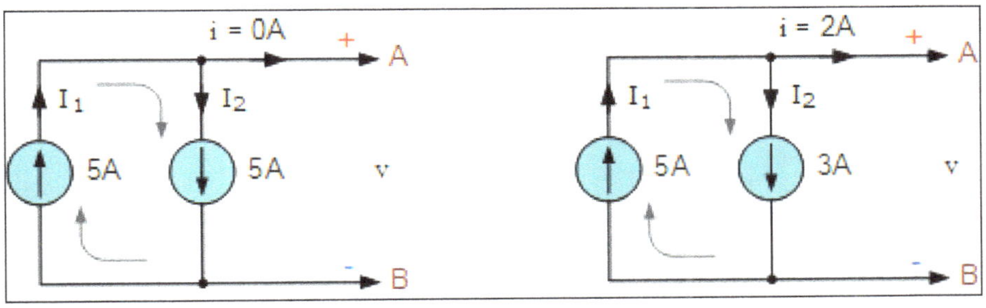

Here, as the two current sources are connected in opposite directions (indicated by their arrows), the two currents subtract from each other as the two provide a closed-loop path for a circulating current complying with Kirchoff's Current Law, KCL. So for example, two current sources of 5 amps each would result in zero output as 5A -5A = 0A. Likewise, if the two currents are of different values, 5A and 3A, then the output will be the subtracted value with the smaller current subtracted from the larger current. Resulting in a I_T of 5 - 3 = 2A.

Ideal current sources can be connected together in parallel to form parallel-aiding or parallel-opposing current sources. What is not allowed or is not best practice for circuit analysis, is connecting together ideal current sources in series combinations.

Current Sources in Series

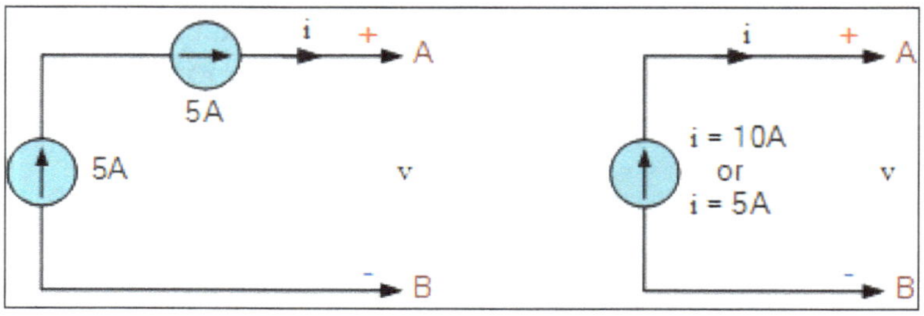

Current sources are not allowed to be connected together in series, either of the same value or ones with different values. Here in this example, two current sources of 5 amps each are connected together in series, but what is the resulting current value. Is it equal to one source of 5 amps, or is it equal to the addition of the two sources, that is 10 amps. Then series connected current sources add an unknown factor into circuit analysis, which is not good.

Also, another reason why series connected sources are not allowed for circuit analysis techniques is that they may not supply the same current in the same direction. Series-aiding or series-opposing currents do not exist for ideal current sources.

Current Source Examples

1. Two current sources of 250 milli-amps and 150 milli-amps respectively are connected together in a parallel-aiding configuration to supply a connected load of 20 ohms. Calculate the voltage drop across the load and the power dissipated. Draw the circuit.

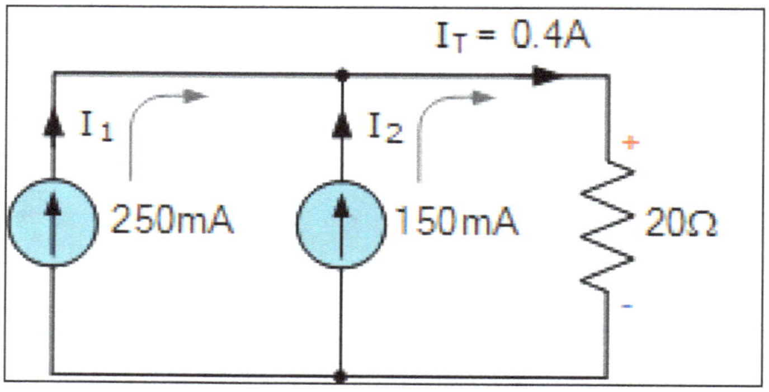

$$I_T = I_1 = I_2 = 0.25 + 0.15 = 0.4\,\text{A}$$
$$V_R = I_T \times R = 0.4 \times 20 = 8\text{V}$$
$$P_R = I_T^2 \times R = 0.4^2 \times 20 = 3.2\text{W}$$

Then, I_T = 0.4A or 400mA, V_R = 8V, and P_R = 3.2W

Practical Current Source

An ideal constant current source can supply the same amount of current indefinitely regardless of the voltage across its terminals, thus making it an independent source. This therefore implies that the current source has an infinite internal resistance, (R = ∞). This idea works well for circuit analysis techniques, but in the real world current sources behave a little differently as practical current sources always have an internal resistance, no matter how large (usually in the mega-ohms range), causing the generated source to vary somewhat with the load.

A practical or non-ideal current source can be represented as an ideal source with an internal resistance connected across it. The internal resistance (R_P) produces the same effect as a resistance connected in parallel (shunt) with the current source as shown. Remember that circuit elements in parallel have exactly the same voltage drop across them.

Ideal and Practical Current Source

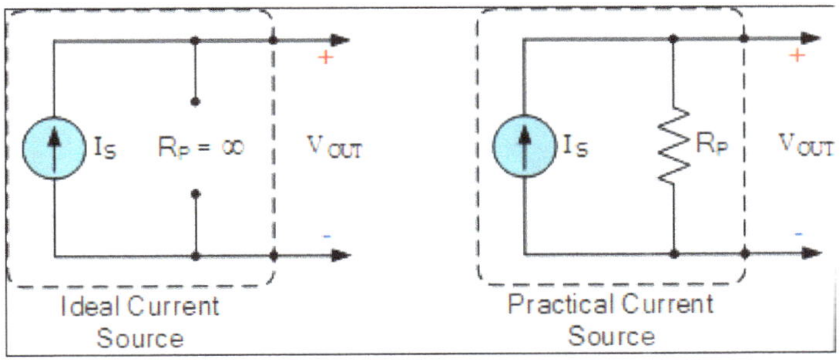

A practical current source closely resembles that of a Norton's equivalent circuit as Norton's theorem states that "any linear dc network can be replaced by an equivalent circuit consisting of a constant-current source, I_S in parallel with a resistor, R_P "if this parallel resistance is very low, $R_P = 0$, the current source is short-circuited. When the parallel resistance is very high or infinite, $R_P \approx \infty$, the current source can be modelled as ideal.

An ideal current source plots a horizontal line on the I-V characteristic .However as practical current sources have an internal source resistance, this takes some of the current so the characteristic of this practical source is not flat and horizontal but will reduce as the current is now splitting into two parts, with one part of the current flowing into the parallel resistance, R_P and the other part of the current flowing straight to the output terminals.

Ohms law tells us that when a current, (i) flows through a resistance, (R) a voltage drop is produce across the same resistance. The value of this voltage drop will be given as $i*R_P$. Then V_{OUT} will be equal to the voltage drop across the resistor with no load attached. We remember that for an ideal source current, R_P is infinite as there is no internal resistance, therefore the terminal voltage will be zero as there is no voltage drop.

The sum of the current around the loop given by Kirchoff's current law, KCL is: $I_{OUT} = I_S - V_S/R_P$. This equation can be plotted to give the I-V characteristics of the output current. It is given as a straight line with a slope $-R_P$ which intersects the vertical voltage axis at the same point as I_S when the source is ideal as shown.

Practical Current Source Characteristics

Therefore, all ideal current sources will have a straight line I-V characteristic but non-ideal or real practical current sources will have an I-V characteristic that is slightly angled down by an amount equal to V_{OUT}/R_P where R_P is the internal source resistance.

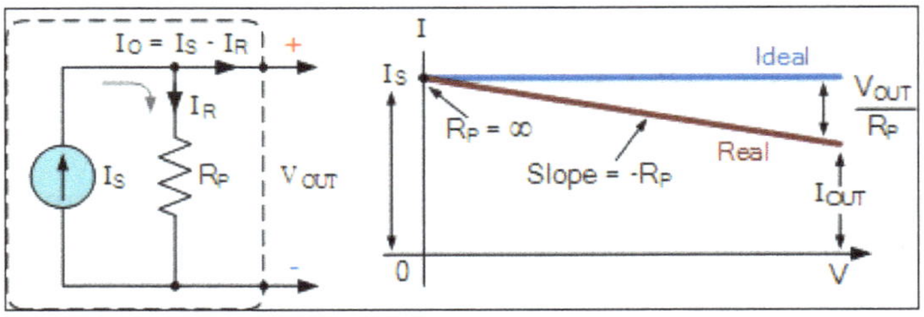

A practical current source consists of a 3A ideal current source which has an internal resistance of 500 Ohms. With no-load attached, calculate the current sources open-circuit terminal voltage and the no-load power absorbed by the internal resistor.

1. No-load values:

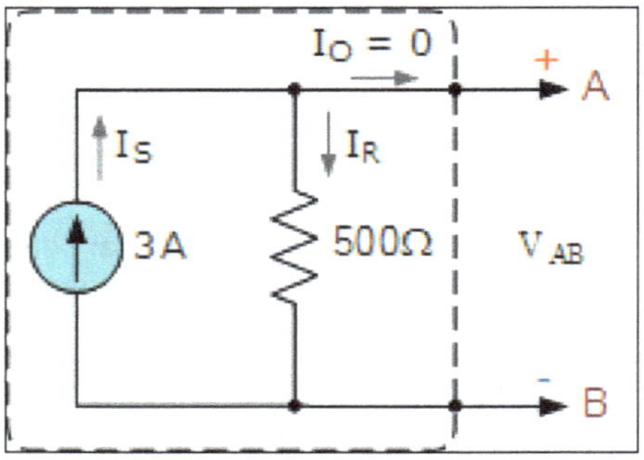

$$I_S = I_R = 3A, RP = 500\Omega$$
$$V_{AB} = V_{RP} = I_S \times R_P$$
$$\therefore V_{RP} = 3 \times 500 = 1500V \text{ Or } 1.5KV$$
$$P_R = I_S^2 \times R_P$$
$$P_R = 3^2 \times 500 = 4500W \text{ or } 4.5W$$

Then the open circuit voltage across the internal source resistance and terminals A and B (V_{AB}) is calculated at 1500 volts.

If a 250 Ohm load resistor is connected to the terminals of the same practical current source, calculate the current through each resistance, the power absorbed by each resistance and the voltage drop across the load resistor. Draw the circuit.

2. Data given with load connected: I_S = 3A, R_P = 500Ω and R_L = 250Ω

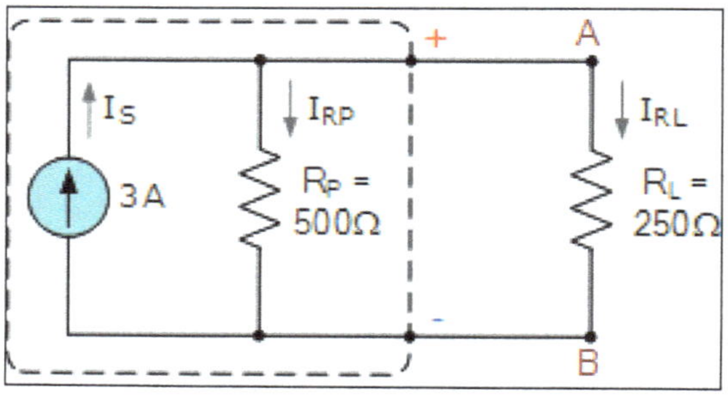

3. To find the currents in each resistive branch, we can use the current-division rule.

$$I_{RP} = \frac{R_L}{R_P + R_L} \times I_S \frac{250}{500 \times 250} \times 3 = 1A$$

$$I_{RL} = \frac{R_P}{R_L + R_P} \times I_S \frac{250}{500 \times 250} \times 3 = 2A$$

$$\therefore I_{RP} + I_{RL} = 1 + 2 = 3A = I_S$$

4. The power absorbed by each resistor is given as:

$$P_{RP} = I_{RP}^2 \times R_P = 1^2 \times 500 = 500W$$

$$P_{RL} = I_{RL}^2 \times R_L = 2^2 \times 500 = 1000W$$

5. Then the voltage drop across the load resistor, R_L is given as:

$$V_{AB} = V_S + R_T$$

$$R_T = \frac{R_P \times R_L}{R_P + R_L} = \frac{500 \times 250}{500 \times 250} = 166.7\Omega$$

$$\therefore V_{AB} = 3 \times 166.7 = 500V$$

We can see that the terminal voltage of an open-circuited practical current source can be very high it will produce whatever voltage is needed, 1500 volts in this example, to supply the specified current. In theory, this terminal voltage can be infinite as the source attempts to deliver the rated current. Connecting a load across its terminals will reduce the voltage, 500 volts in this example, as now the current has somewhere to go and for a constant current source, the terminal voltage is directly proportional to the load resistance.

In the case of non-ideal current sources that each have an internal resistance, the total internal resistance (or impedance) will be the result of combining them together in parallel, exactly the same as for resistors in parallel.

Dependent Current Source

An ideal current source provides a specified amount of current completely independent of the voltage across it and as such will produce whatever voltage is necessary to maintain the required current. This then makes it completely independent of the circuit to which it is connected to resulting in it being called an ideal independent current source.

A controlled or dependent current source on the other hand changes its available current depending upon the voltage across, or the current through, some other element connected to the circuit. In other words, the output of a dependent current source is controlled by another voltage or current.

Dependent current sources behave similar to the current sources we have looked at so far, both ideal (independent) and practical. The difference this time is that a dependent current source can be controlled by an input voltage or current. A current source that depends on a voltage input is generally referred to as a Voltage Controlled Current Source or VCCS. A current source that depends on a current input is generally referred to as a Current Controlled Current Source or CCCS. Generally, an ideal current dependent source, either voltage or current controlled is designated by a diamond-shaped symbol where an arrow indicates the direction of the current, i as shown.

Dependent Current Source Symbols

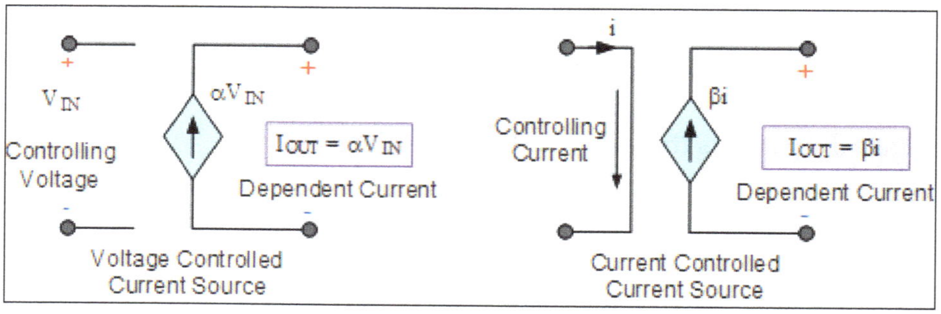

An ideal dependent voltage-controlled current source, VCCS, maintains an output current, I_{OUT} that is proportional to the controlling input voltage, V_{IN}. In other words, the output current "depends" on the value of input voltage making it a dependent current source.

Then the VCCS output current is defined by the following equation: $I_{OUT} = \alpha V_{IN}$. This multiplying constant *α* (alpha) has the SI units of mhos, ℧ (an inverted Ohms sign) because $\alpha = I_{OUT}/V_{IN}$, and its units will therefore be amperes/volt.

An ideal dependent current-controlled current source, CCCS, maintains an output current that is proportional to a controlling input current. Then the output current "depends" on the value of the input current, again making it a dependent current source.

As a controlling current, I_{IN} determines the magnitude of the output current, I_{OUT} times the magnification constant β (beta), the output current for a CCCS element is determined by the following equation: $I_{OUT} = \beta I_{IN}$. The multiplying constant *β* is a dimensionless scaling factor as $\beta = I_{OUT}/I_{IN}$, so therefore its units would be amperes/amperes.

Filter

An electric filter is usually a frequency-selective network that passes a specified band of frequencies and blocks or attenuates signals of frequencies outside this band.

Depending on the types of elements used in their construction, filters may be passive or active. A passive filter is built with passive components such as resistors, capacitors and inductors. Active filters, on the other hand, make use of transistors or op-amps (providing voltage amplification, and signal isolation or buffering) in addition to resistors and capacitors. The type of the elements used dictates the operating frequency range of the filter.

According to the operating frequency range, the filters may be classified as audio-frequency (AF) or radio-frequency (RF) filters.

Filters may also be classified as (i) low-pass, (ii) high-pass (iii) band-pass (iv) band-stop. The filter circuit may be so designed that some frequencies are passed from the input to the output of the filter with very little attenuation while others are greatly attenuated.

Figure below shown the frequency responses of the four types (mentioned above) of filters. These are ideal responses and cannot be achieved in actual practice.

- Ideal Low Pass Filter: A filters that provides a constant output from dc up to a cut-off frequency fc and then passes no signal above that frequency is called an ideal low-pass filter. The ideal response of a low-pass filter is illustrate in fig.(a) The voltage gain (the ration of output voltage and input voltage i.e.(V_{out}/V_{in}) is constant over a frequency range from zero to cut off frequency, fc. The fc output of any signal having a frequency exceeding fc will be attenuated i.e there will be no output voltage for frequencies exceeding cut-off frequency fc. Hence output will be available faithfully from 0 to fc with constant gain, and is 0 from fc onward. The frequencies between 0 and fc are, therefore, called the passband frequencies, while the range of frequencies, those beyond fc, that are attenuated include the stopband frequencies. The bandwidth (BW) is, therefore fc.

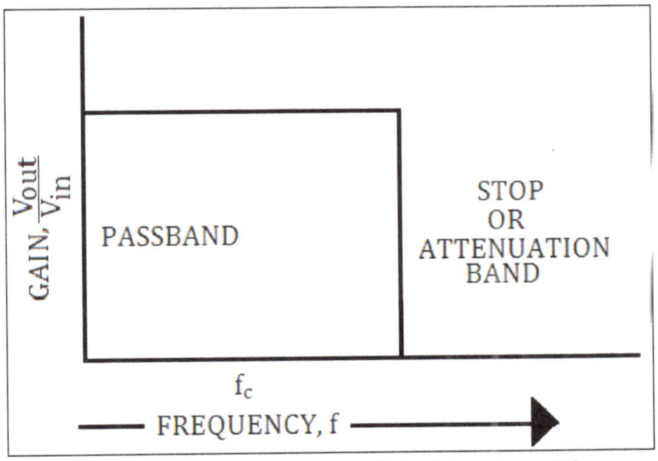

Figure: Frequency Response of Ideal Low Pass Filter

- Ideal High Pass Filter: A filter that provides or passes signals above a cut-off frequency fc is a high-pass filter, as idealized in figure(below) frequency fc, called the cut-off frequency, and above this frequency, the gain is constant, as illustrated in fig. Thus signal of any frequency beyond fc is faithfully reproduced with a constant gain, and frequencies from 0 to fc will be attenuated.

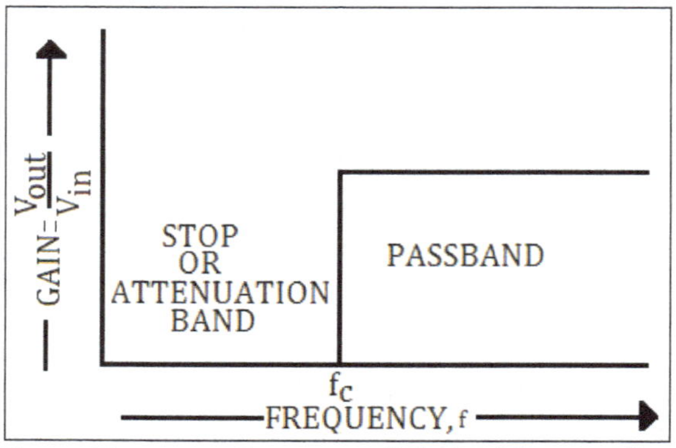

Figure: Frequency Response of Ideal High Pass Filter

- Ideal Band-Pass Filter: When the filter circuit passes signals that are above one cut-off frequency and below a second cut-off frequency, it is called a band-pass filter, as idealized Fig.. Thus a band-pass filter has a passband between two cut-off frequencies f_{c2} and f_{f1} ,where f_{c2} and f_{c1} and two stop-bands: $0 < f < f_{c1}$ and $f > f_{c2}$ and $f > f_{f2}$ are lower and higher cut-off frequencies respectively.

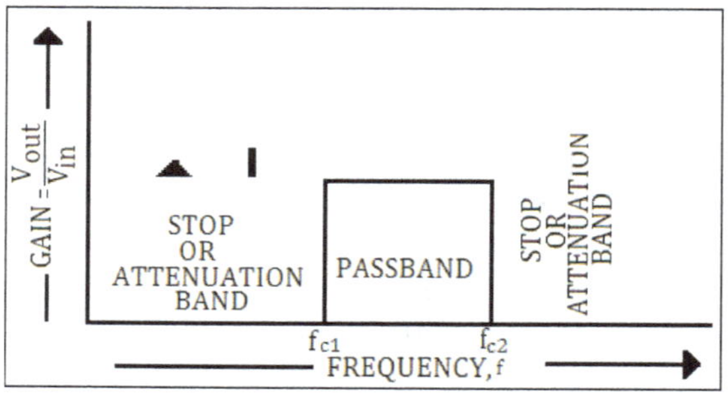

Figure: Frequency Response of Ideal Band Pass Filter

- Ideal Band-Stop Filter: The band-stop or band-reject filter performs exactly opposite to the band-pass i.e. it has a bandstop between two cut-off frequencies f_{c1} and f_{c2} two pass-bands: $0 < f < f_{c1}$ and $f > f_{c2}$. The ideal response of a band-stop filter is illustrates in Fig. This is also called a band-elimation or notch filter.

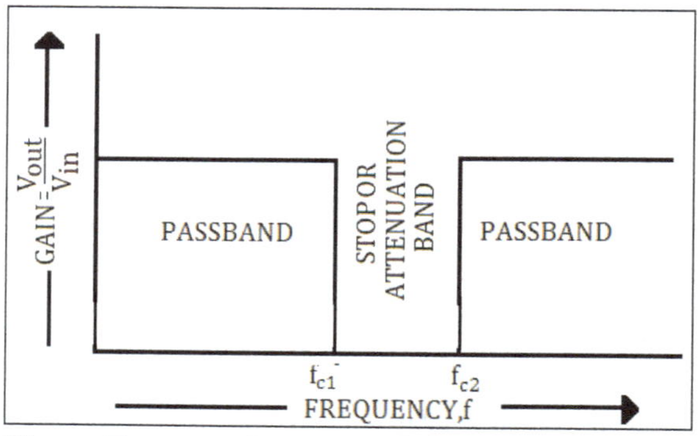

Figure: Frequency Response of Ideal Band stop Filter

Every practical filter fails to satisfy the ideal conditions of zero attenuation over passband and infinity attenuation over stop band. This results due to reflections from terminal impedance inequalities and due to losses cause by imperfect conductor and dielectrics.

The filter discussed above have ideal characteristics and a sharp cut-off but unfortunately, filter response is not practical because linear networks cannot produce the discontinuities. However, it is possible to obtain a practical response that approximate the ideal response by using special design techniques, as well as precision component values and high-speed op-amps.

General Requirement of a Filter: Filters are required to have zero attenuation in the pass band. Hence, it is necessary that a filter be constructed from purely reactive element namely L and C and no resistance should be included in the circuit. This requirement is never met fully since every inductor always possesses some resistance.

Factor Characterizing a Filter

1. Characteristic Impedance Z_0: This should be so selected that the filter may fit into a given transmission line or between two pieces of equipment's.
2. Pass Band: Filter should have very low (ideally zero) attenuation in the pass band and very high (ideally infinity) attenuation in the stop band or attenuation band.
3. Cutoff Frequency: This frequency forms the boundary between the pass band and the stop band.

References

- Resistors: sparkfun.com, Retrieved 7 April, 2019
- Resistors: explainthatstuff.com, Retrieved 18 July, 2019
- Resistor: electronics-tutorials.ws, Retrieved 29 January, 2019
- What-is-a-resistor: resistorguide.com, Retrieved 26 March, 2019
- What-is-a-capacitor, circuit-playground-c-is-for-capacitor: adafruit.com, Retrieved 16 May, 2019
- Capacitor: electronics-tutorials.ws, Retrieved 6 February, 2019
- Introduction-to-inductors: components101.com, Retrieved 8 March, 2019
- Transistors-basics-types-baising-modes: elprocus.com, Retrieved 18 August, 2019
- Diode-working-principle-and-types-of-diode: electrical4u.com, Retrieved 9 January, 2019
- Diodes: engineersgarage.com, Retrieved 21 March, 2019
- Voltage-source, dccircuits: electronics-tutorials.ws, Retrieved 23 May, 2019
- Current-source, dccircuits: electronics-tutorials.ws, Retrieved 3 February, 2019
- Electric-filter-types-of-filter: electronicspani.com, Retrieved 27 June, 2019

3

Laws and Theorems of Circuits

There are a number of theorems and laws pertaining to circuits such as Ohm's law, Norton's theorem and Superposition Theorem. Ohm's law provides the relationship between current, resistance and voltage. The diverse applications of these laws and theorems in different scenarios have been thoroughly discussed in this chapter.

Ohm's Law

Ohm's law states that the voltage or potential difference between two points is directly proportional to the current or electricity passing through the resistance, and inversely proportional to the resistance of the circuit. The formula for Ohm's law is V=I/R. This relationship between current, voltage and relationship was discovered by German scientist Georg Simon Ohm.

Most basic components of electricity are voltage, current, and resistance. Ohm's law shows a simple relation between these three quantities. Ohm's law states that the current through a conductor between two points is directly proportional to the voltage across the two points.

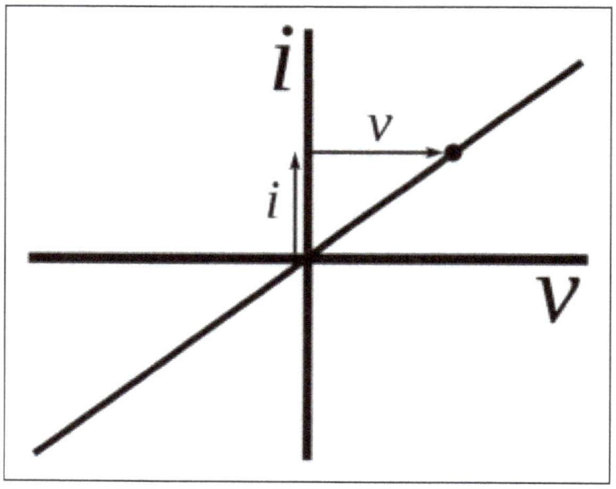

Ohm's Law Formula

Voltage= Current× Resistance

$$V = I \times R$$

V= voltage, I= current and R= resistance

The SI unit of resistance is ohms and is denoted by Ω

This law is one of the most basic laws of electricity. It helps to calculate the power, efficiency, current, voltage, and resistance of an element of an electrical circuit.

Applications of Ohm's Law

Ohm's law helps us in determining either voltage current or impedance or resistance of a linear electric circuit when the other two quantities are known to us. It also makes power calculation simpler.

Relationship between Current-voltage

In order to establish the current-voltage relationship, the ratio V/I remains constant for a given resistance, therefore a graph between the potential difference(V) and the current (I) must be a straight line.

Finding the Values of Resistance

It is the constant ratio that gives the unknown values of resistance,

$$\frac{V}{1}=R$$

For a wire of uniform cross-section, the resistance depends on the length l and the area of cross-section A. It also depends on the temperature of the conductor. At a given temperature the resistance,

$$R=\rho\frac{1}{A}$$

Where ρ is the specific resistance or resistivity and is characteristic of the material of wire. The specific resistance or resistivity of the material of the wire is,

$$\rho=\frac{RA}{1}$$

If 'r' is the radius of the wire, then the cross-sectional area, $A = \pi r^2$. Then the specific resistance or resistivity of the material of the wire is,

$$\rho=\frac{\pi r^2 R}{1}$$

Limitations of Ohms Law

1. Ohm's law is not applicable to unilateral networks. Unilateral networks allow the current to flow in one direction. Such types of network consist of elements like a diode, transistor, etc.
2. Ohm's law is also not applicable to non – linear elements. Non-linear elements are those which do not have current exactly proportional to the applied voltage that means the resistance value of those elements changes for different values of voltage and current. Examples of non – linear elements are the thyristor.

Kirchhoff's Voltage Law (KVL)

The principle known as Kirchhoff's Voltage Law can be stated as such:

"The algebraic sum of all voltages in a loop must equal zero".

Demonstrating Kirchhoff's Voltage Law in a Series Circuit

Let's take another look at our example series circuit, this time numbering the points in the circuit for voltage reference:

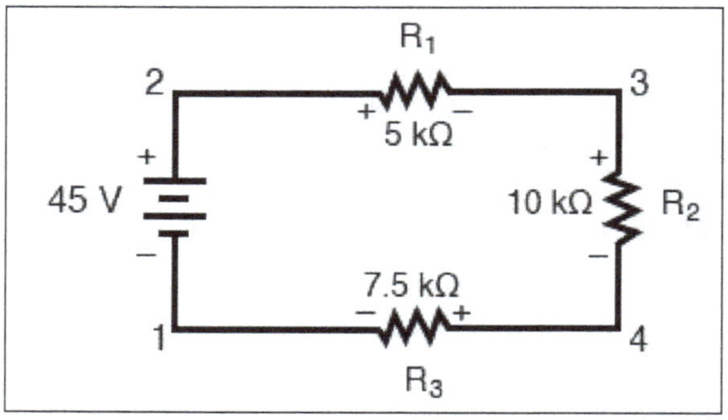

If we were to connect a voltmeter between points 2 and 1, red test lead to point 2 and the black test lead to point 1, the meter would register +45 volts. Typically the "+" sign is not shown but rather implied, for positive readings in digital meter displays, the polarity of the voltage reading is very important and,

$$E_{2\text{-}1} = +45\text{V}$$

When a voltage is specified with a double subscript (the characters "2-1" in the notation "$E_{2\text{-}1}$"), it means the voltage at the first point (2) as measured in reference to the second point (1). A voltage specified as "E_{cd}" would mean the voltage as indicated by a digital meter with the red test lead on point "c" and the black test lead on point "d": the voltage at "c" in reference to "d".

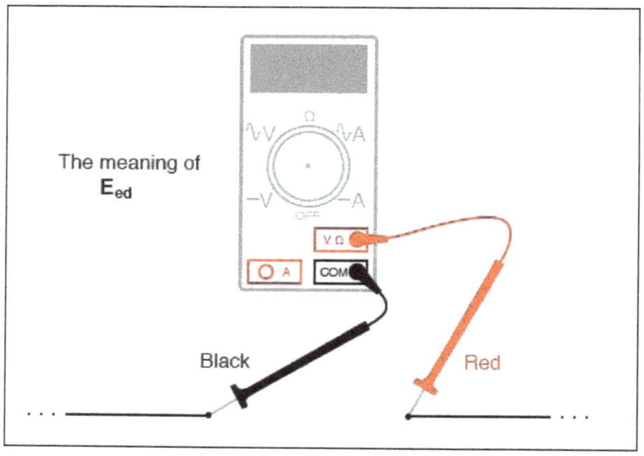

If we were to take that same voltmeter and measure the voltage drop across each resistor, stepping around the circuit in a clockwise direction with the red test lead of our meter on the point ahead and the black test lead on the point behind, we would obtain the following readings:

$E_{3\text{-}2} = +10V$

$E_{4\text{-}3} = -20V$

$E_{1\text{-}4} = -15V$

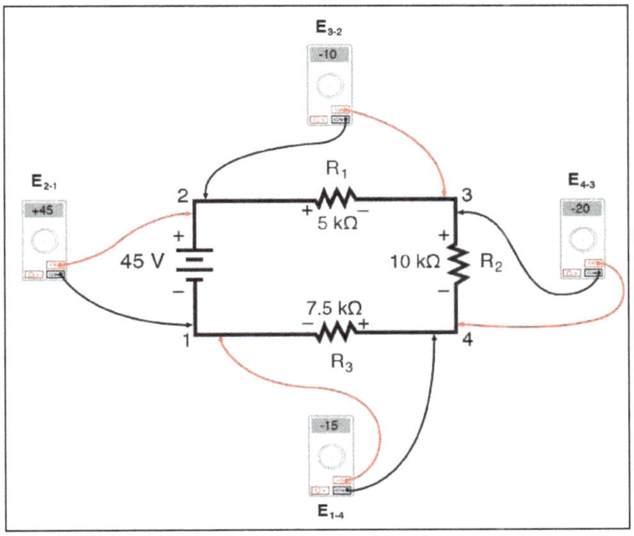

We should already be familiar with the general principle for series circuits stating that individual voltage drops add up to the total applied voltage, but measuring voltage drops in this manner and paying attention to the polarity (mathematical sign) of the readings reveals another facet of this principle: that the voltages measured as such all add up to zero:

$E_{2\text{-}1} = +45V$	Voltage from point 3 to point 2
$E_{3\text{-}2} = -10V$	Voltage from point 4 to point 3
$E_{4\text{-}3} = -20V$	Voltage from point 5 to point 4
$+E_{1\text{-}4} = -15V$	Voltage from point 6 to point 5
$= 0V$	

In the above example, the loop was formed by the following points in this order: 1-2-3-4-1. It doesn't matter which point we start at or which direction we proceed in tracing the loop; the voltage sum will still equal zero. To demonstrate, we can tally up the voltages in loop 3-2-1-4-3 of the same circuit:

$E_{2\text{-}3} = +10V$	Voltage from point 3 to point 2
$E_{1\text{-}2} = -45V$	Voltage from point 4 to point 3
$E_{4\text{-}1} = +15V$	Voltage from point 5 to point 4
$+E_{3\text{-}4} = +20V$	Voltage from point 6 to point 5
$= 0v$	

This may make more sense if we re-draw our example series circuit so that all components are represented in a straight line:

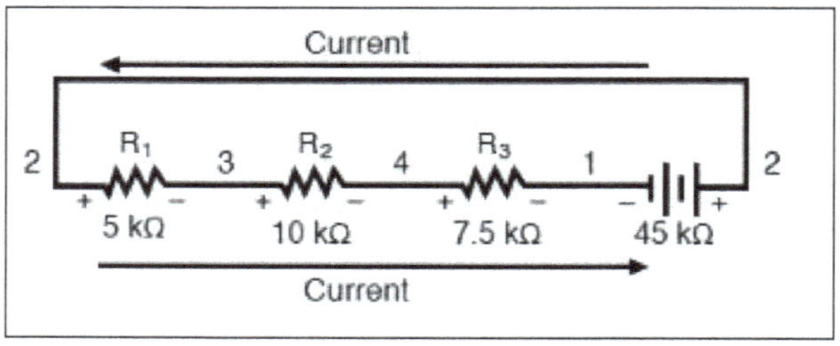

It's still the same series circuit, just with the components arranged in a different form. Notice the polarities of the resistor voltage drops with respect to the battery: the battery's voltage is negative on the left and positive on the right, whereas all the resistor voltage drops are oriented the other way: positive on the left and negative on the right. This is because the resistors are *resisting* the flow of electric charge being pushed by the battery. In other words, the "push" exerted by the resistors *against* the flow of electric charge *must* be in a direction opposite the source of electromotive force.

Here we see what a digital voltmeter would indicate across each component in this circuit, black lead on the left and red lead on the right, as laid out in horizontal fashion:

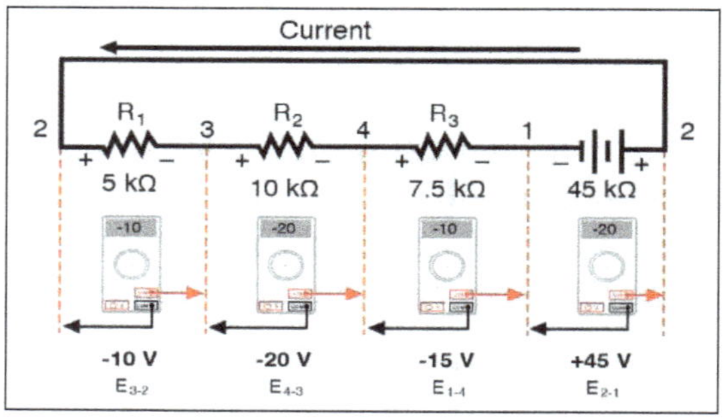

If we were to take that same voltmeter and read voltage across combinations of components, starting with the only R_1 on the left and progressing across the whole string of components, we will see how the voltages add algebraically (to zero):

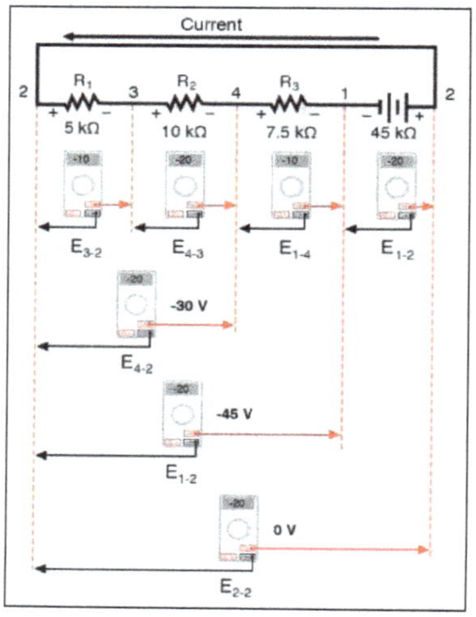

The fact that series voltages add up should be no mystery, but we notice that the *polarity* of these voltages makes a lot of difference in how the figures add. While reading voltage across R_1—R_2, and R_1—R_2—R_3, we see how the voltages measure successively larger (albeit negative) magnitudes, because the polarities of the individual voltage drops are in the same orientation (positive left, negative right). The sum of the voltage drops across R_1, R_2, and R_3 equals 45 volts, which is the same as the battery's output, except that the battery's polarity is opposite that of the resistor voltage drops (negative left, positive right), so we end up with 0 volts measured across the whole string of components.

That we should end up with exactly 0 volts across the whole string should be no mystery, either. Looking at the circuit, we can see that the far left of the string (left side of R_1: point number 2) is directly connected to the far right of the string (right side of battery: point number 2), as necessary to complete the circuit. Since these two points are directly connected, they are electrically common to each other. And, as such, the voltage between those two electrically common points must be zero.

Demonstrating Kirchhoff's Voltage Law in a Parallel Circuit

Kirchhoff's Voltage Law (sometimes denoted as *KVL* for short) will work for *any* circuit configuration at all, not just simple series.

How it works for this parallel circuit:

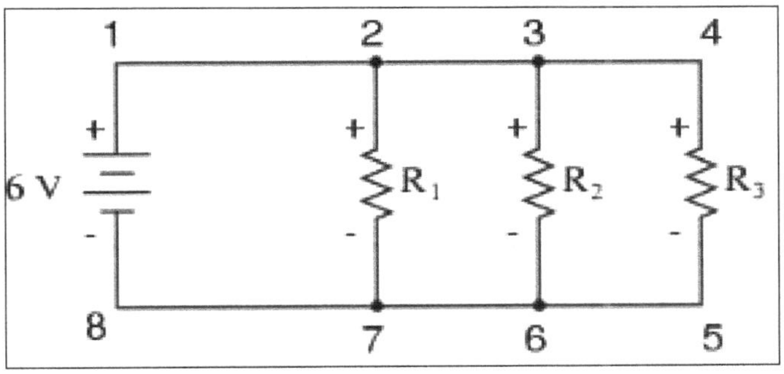

Being a parallel circuit, the voltage across every resistor is the same as the supply voltage: 6 volts. Tallying up voltages around loop 2-3-4-5-6-7-2, we get:

$E_{3-2} = 0V$ voltage from point 3 to point 2

$E_{4-3} = 0V$ voltage from point 4 to point 3

$E_{5-4} = +6V$ voltage from point 5 to point 4

$E_{6-5} = 0V$ voltage from point 6 to point 5

$E_{7-6} = 0V$ voltage from point 7 to point 6

$+E_{2-7} = +6V$ voltage from point 2 to point 7

$E_{2-2} = 0v$

The Validity of Kirchhoff's Voltage Law and Regardless of Circuit Topology

The fact that this circuit is parallel instead of series has nothing to do with the validity of Kirchhoff's

Voltage Law. For that matter, the circuit could be a "black box"—its component configuration completely hidden from our view, with only a set of exposed terminals for us to measure the voltage between—and KVL would still hold true,

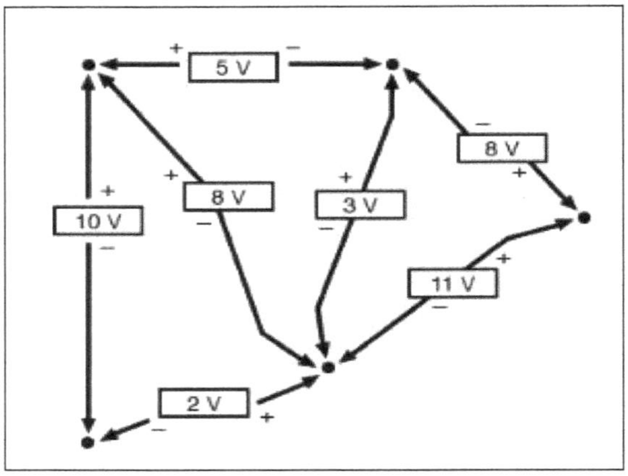

Try any order of steps from any terminal in the above diagram, stepping around back to the original terminal, and you'll find that the algebraic sum of the voltages always equals zero.

The "loop" we trace for KVL doesn't even have to be a real current path in the closed-circuit sense of the word. All we have to do to comply with KVL is to begin and end at the same point in the circuit, tallying voltage drops and polarities as we go between the next and the last point. Consider this absurd example, tracing "loop" 2-3-6-3-2 in the same parallel resistor circuit:

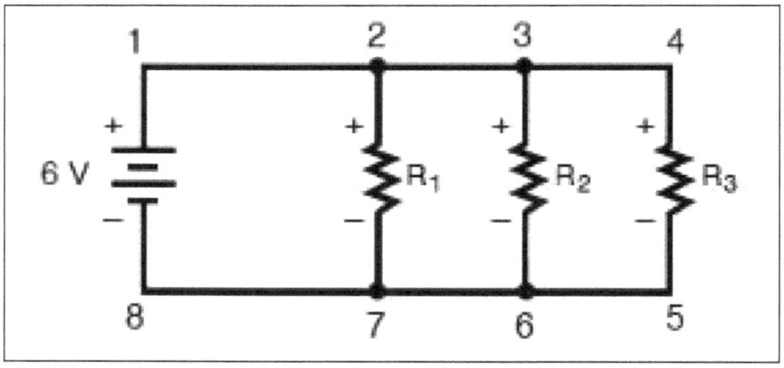

$$
\begin{array}{ll}
E_{3\text{-}2} = 0v & \text{voltage from point 3 to point 2} \\
E_{6\text{-}3} = -6v & \text{voltage from point 6 to point 3} \\
E_{3\text{-}6} = +6v & \text{voltage from point 3 to point 6} \\
+E_{2\text{-}3} = 0v & \text{voltage from point 2 to point 3} \\
\hline
E_{2\text{-}2} = 0v &
\end{array}
$$

Using Kirchhoff's Voltage Law in a Complex Circuit

KVL can be used to determine an unknown voltage in a complex circuit, where all other voltages around a particular "loop" are known. Take the following complex circuit (actually two series circuits joined by a single wire at the bottom) as an example:

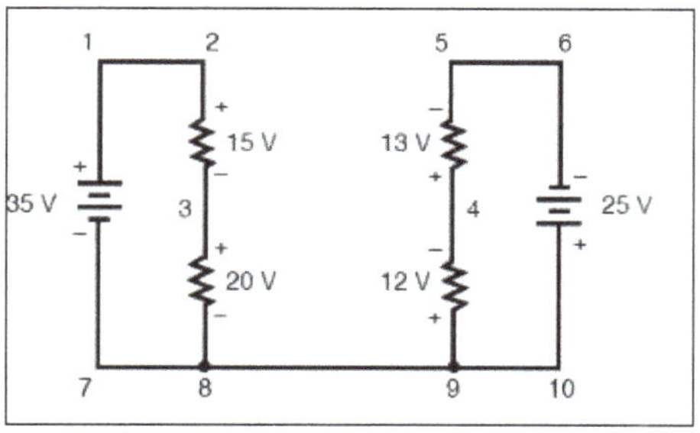

The two series circuits share a common wire between them (wire 7-8-9-10), making voltage measurements *between* the two circuits possible. If we wanted to determine the voltage between points 4 and 3, we could set up a KVL equation with the voltage between those points as the unknown:

$$E_{4-3}, + E_{9-4} + E_{8-9} + E_{3-8} = 0$$

$$E_{4-3} + 12 + 0 + 20 = 0$$

$$E_{4-3} + 32 = 0$$

$$E_{4-3} = -32\text{V}$$

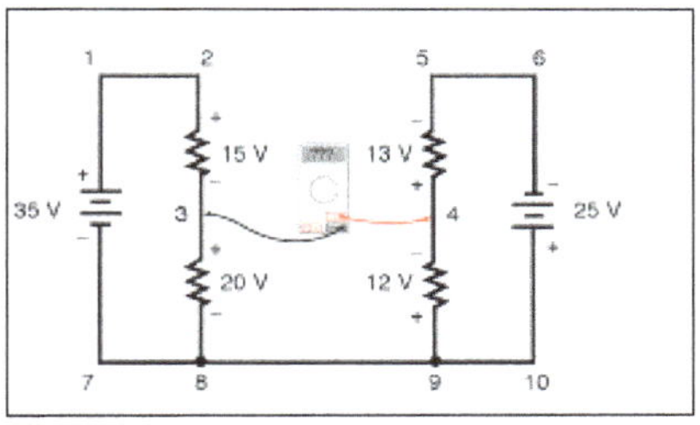

Measuring voltage from point 4 to point 3 (unknown amount) E_{4-3}

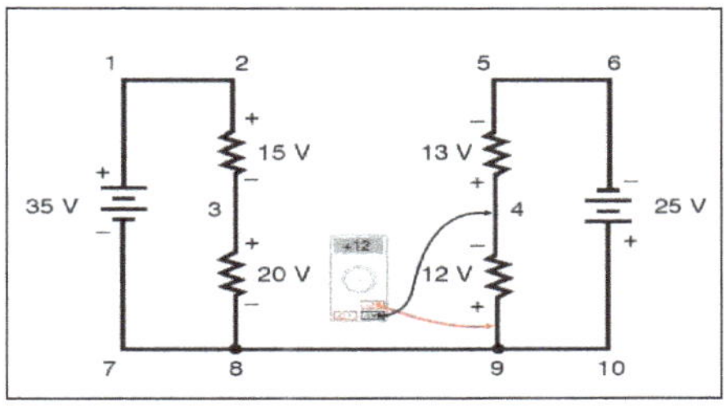

Measuring voltage from point 9 to point 4 (+12volts) E_{4-3}= + 12

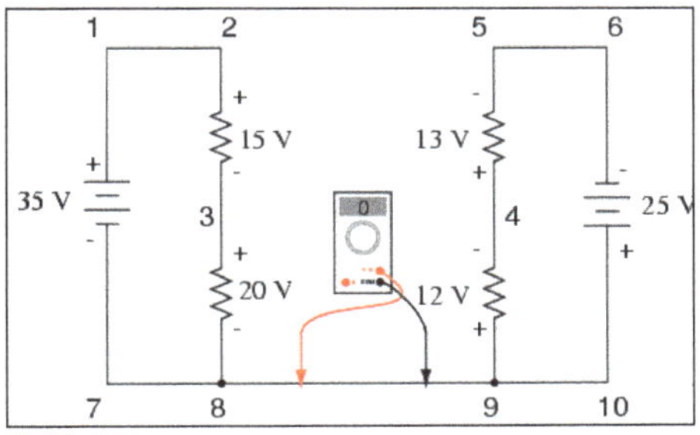

Measuring voltage from point 9 to point 4 (+20volts) E_{4-3} + 20 + 0

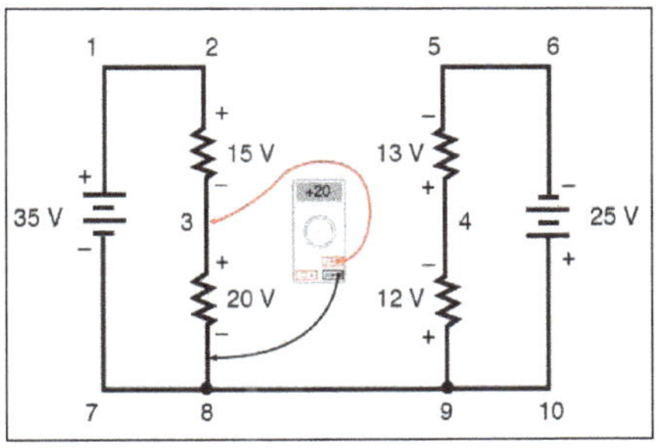

Measuring voltage from point 9 to point 4 (+20volts) E_{4-3}= + 12 + 0 + 20 = 0

Stepping around the loop 3-4-9-8-3, we write the voltage drop figures as a digital voltmeter would register them, measuring with the red test lead on the point ahead and black test lead on the point behind as we progress around the loop. Therefore, the voltage from point 9 to point 4 is a positive (+) 12 volts because the “red lead” is on point 9 and the “black lead” is on point 4. The voltage from point 3 to point 8 is a positive (+) 20 volts because the “red lead” is on point 3 and the “black lead” is on point 8. The voltage from point 8 to point 9 is zero, because those two points are electrically common.

The voltage from point 4 to point 3 is a negative (-) 32 volts, telling us that point 3 is actually positive with respect to point 4, precisely what a digital voltmeter would indicate with the red lead on point 4 and the black lead on point 3:

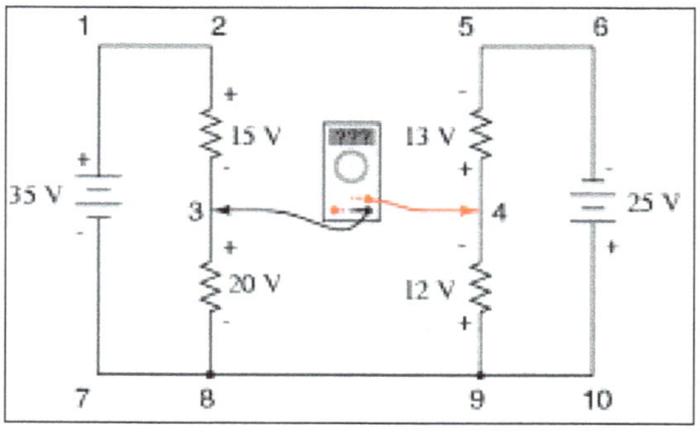

$E_{4\text{-}3} = +32$

In other words, the initial placement of our "meter leads" in this KVL problem was "backward." Had we generated our KVL equation starting with $E_{3\text{-}4}$ instead of $E_{4\text{-}3}$, stepping around the same loop with the opposite meter lead orientation, the final answer would have been $E_{3\text{-}4} = +32$ volts:

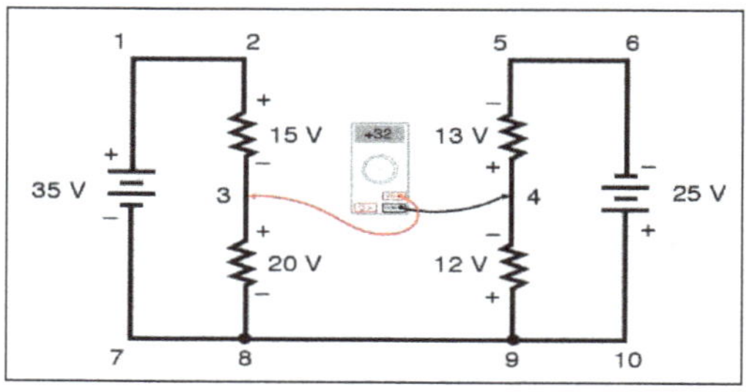

$E_{4\text{-}3} = +32$

It is important to realize that neither approach is "wrong." In both cases, we arrive at the correct assessment of voltage between the two points, 3 and 4: point 3 is positive with respect to point 4, and the voltage between them is 32 volts.

Kirchhoff's Current Law (KCL)

Kirchhoff's Current Law (KCL) is Kirchhoff's first law that deals with the conservation of charge entering and leaving a junction.

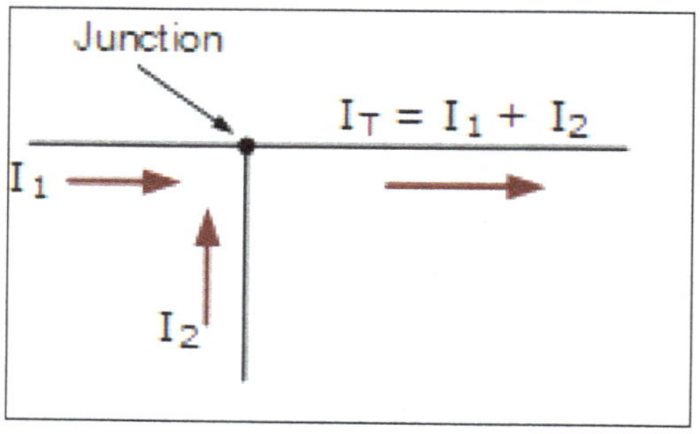

To determine the amount or magnitude of the electrical current flowing around an electrical or electronic circuit, we need to use certain laws or rules that allow us to write down these currents in the form of an equation. The network equations used are those according to Kirchhoff's laws, and as we are dealing with circuit currents, we will be looking at Kirchhoff's current law, (KCL).

Gustav Kirchhoff's Current Law is one of the fundamental laws used for circuit analysis. His current law states that for a parallel path the total current entering a circuit junction is exactly equal to the total current leaving the same junction. This is because it has no other place to go as no charge is lost.

In other words the algebraic sum of ALL the currents entering and leaving a junction must be equal to zero as: $\Sigma I_{IN} = \Sigma I_{OUT}$.

This idea by Kirchhoff is commonly known as the Conservation of Charge, as the current is conserved around the junction with no loss of current. Let's look at a simple example of Kirchhoff's current law (KCL) when applied to a single junction.

A Single Junction

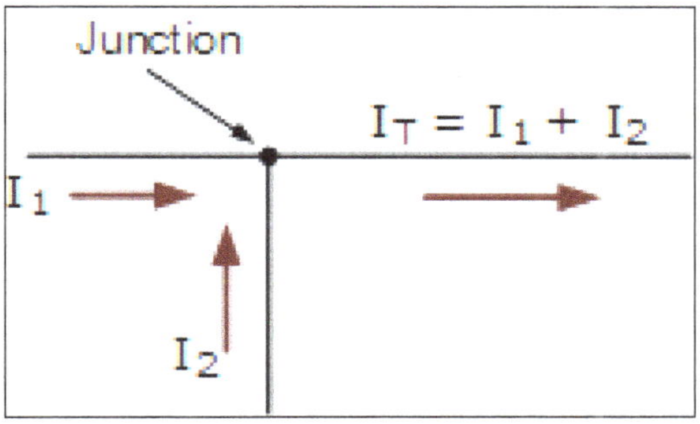

Here in this simple single junction example, the current I_T leaving the junction is the algebraic sum of the two currents, I_1 and I_2 entering the same junction. That is $I_T = I_1 + I_2$. We could also write this correctly as the algebraic sum of: $I_T - (I_1 + I_2) = 0$.

So if I_1 equals 3 amperes and I_2 is equal to 2 amperes, then the total current, I_T leaving the junction will be 3 + 2 = 5 amperes, and we can use this basic law for any number of junctions or nodes as the sum of the currents both entering and leaving will be the same.

Also, if we reversed the directions of the currents, the resulting equations would still hold true for I_1 or I_2. As $I_1 = I_T - I_2 = 5 - 2 = 3$ amps, and $I_2 = I_T - I_1 = 5 - 3 = 2$ amps. Thus we can think of the currents entering the junction as being positive (+), while the ones leaving the junction as being negative (-).

Then we can see that the mathematical sum of the currents either entering or leaving the junction and in whatever direction will always be equal to zero, and this forms the basis of Kirchhoff's Junction Rule, more commonly known as Kirchhoff's Current Law, or (KCL).

Resistors in Parallel

Let's look how we could apply Kirchhoff's current law to resistors in parallel, whether the resistances in those branches are equal or unequal.

Consider the following circuit diagram:

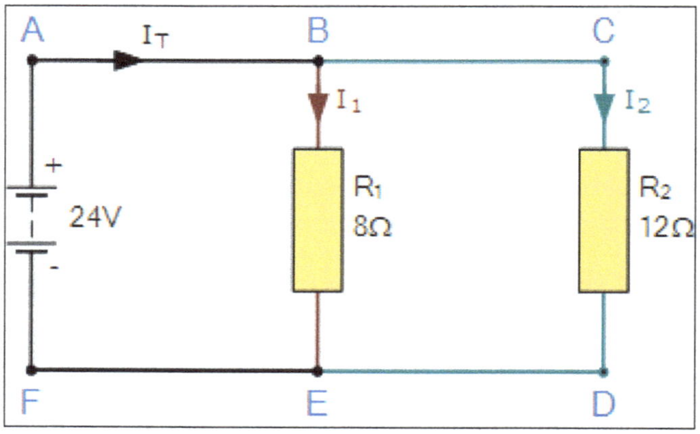

In this simple parallel resistor example there are two distinct junctions for current. Junction one occurs at node B, and junction two occurs at node E. Thus we can use Kirchhoff's Junction Rule for the electrical currents at both of these two distinct junctions, for those currents entering the junction and for those currents flowing leaving the junction.

To start, all the current, I_T leaves the 24 volt supply and arrives at point A and from there it enters node B. Node B is a junction as the current can now split into two distinct directions, with some of the current flowing downwards and through resistor R_1 with the remainder continuing on through resistor R_2 via node C. the currents flowing into and out of a node point are commonly called branch currents.

We can use Ohm's Law to determine the individual branch currents through each resistor as: I = V/R, thus:

For current branch B to E through resistor R_1

$$I_{B\text{-}E} = I_1 = \frac{V}{R_1} = \frac{24}{8} = 3A$$

For current branch C to D through resistor R_2

$$I_{C\text{-}D} = I_2 = \frac{V}{R_2} = \frac{24}{12} = 2A$$

From above we know that Kirchhoff's current law states that the sum of the currents entering a junction must equal the sum of the currents leaving the junction, and in our simple example above, there is one current, I_T going into the junction at node B and two currents leaving the junction, I_1 and I_2.

Since we now know from calculation that the currents leaving the junction at node B is I_1 equals 3 amps and I_2 equals 2 amps, the sum of the currents entering the junction at node B must equal 3 + 2 = 5 amps. Thus $\Sigma_{IN} = I_T = 5$ amperes.

In our example, we have two distinct junctions at node B and node E, thus we can confirm this value for I_T as the two currents recombine again at node E. So, for Kirchhoff's junction rule to hold true, the sum of the currents into point F must equal the sum of the currents flowing out of the junction at node E.

As the two currents entering junction E are 3 amps and 2 amps respectively, the sum of the currents entering point F is therefore: 3 + 2 = 5 amperes. Thus $\Sigma_{IN} = I_T = 5$ amperes and therefore Kirchhoff's current law holds true as this is the same value as the current leaving point A.

Applying KCL to more Complex Circuits

We can use Kirchhoff's current law to find the currents flowing around more complex circuits. We hopefully know by now that the algebraic sum of all the currents at a node (junction point) is equal to zero and with this idea in mind, it is a simple case of determining the currents entering a node and those leaving the node. Consider the circuit below.

Kirchhoff's Current Law Examples

1. In this example there are four distinct junctions for current to either separate or merge together at nodes A, C, E and node F. The supply current I_T separates at node A flowing through resistors R_1 and R_2, recombining at node C before separating again through resistors R_3, R_4 and R_5 and finally recombining once again at node F.

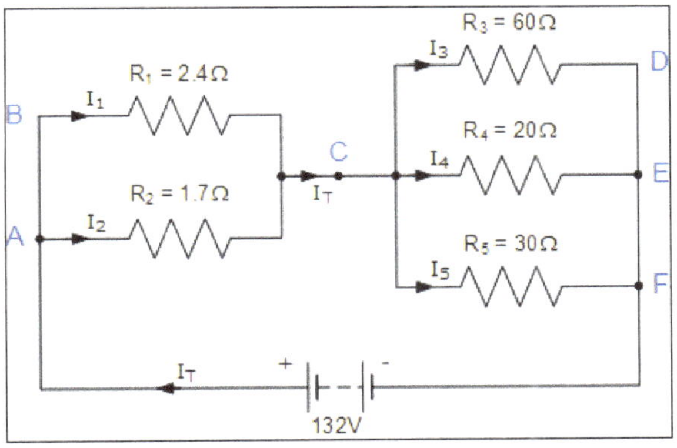

But before we can calculate the individual currents flowing through each resistor branch, we must first calculate the circuits total current, I_T. Ohms law tells us that I = V/R and as we know the value of V, 132 volts, we need to calculate the circuit resistances as follows.

Circuit Resistance R_{AC}

$$\frac{1}{R_{(AC)}} = \frac{1}{R_1} + \frac{1}{R_2} + \frac{1}{2.4} + \frac{1}{1.7}$$

$$\frac{1}{R_{(AC)}} = \frac{1}{1} \quad \therefore R_{(AC)} = 1\Omega$$

Thus the equivalent circuit resistance between nodes A and C is calculated as 1 Ohm.

Circuit Resistance R_{CF}

$$\frac{1}{R_{(CF)}} = \frac{1}{R_3} + \frac{1}{R_4} + \frac{1}{R_5} + \frac{1}{60} + \frac{1}{20} + \frac{1}{30}$$

$$\frac{1}{R_{(CF)}} \quad \frac{1}{0.1} \qquad \therefore R_{(CF)} = 10\Omega$$

Thus the equivalent circuit resistance between nodes C and F is calculated as 10 Ohms. Then the total circuit current, I_T is given as:

$$R_T = R_{(AC)} + R_{(CF)} + 1 + 10 = 11\Omega$$

$$I_T = \frac{V}{R_T} = \frac{132}{11} = 12\,\text{Amps}$$

Kirchhoff's Current Law Equivalent Circuit,

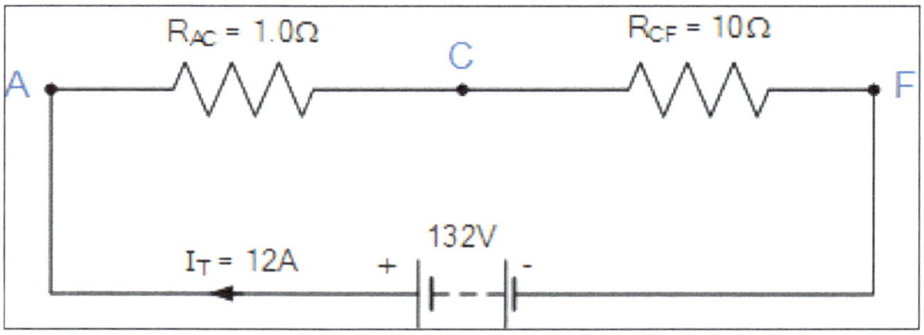

Therefore, V = 132V, $R_{AC} = 1\Omega$, $R_{CF} = 10\Omega$'s and $I_T = 12A$.

Having established the equivalent parallel resistances and supply current, we can now calculate the individual branch currents and confirm using Kirchhoff's junction rule as follows.

$$V_{AC} = I_T \text{ x } R_{Ac} = 12\text{x}1 = 12 \text{ Volts}$$

$$V_{cF} = I_T \text{ x } R_{cF} = 12\text{x}10 = 120 \text{ Volts}$$

$$I_1 = \frac{V_{AC}}{R_1} = \frac{12}{2.4} = 5 \text{ Amps}$$

$$I_2 = \frac{V_{AC}}{R_2} = \frac{12}{1.7} = 7 \text{ Amps}$$

$$I_3 = \frac{V_{CF}}{R_3} = \frac{120}{60} = 2 \text{ Amps}$$

$$I_4 = \frac{V_{CF}}{R_4} = \frac{120}{20} = 6 \text{ Amps}$$

$$I_5 = \frac{V_{CF}}{R_5} = \frac{120}{30} = 4 \text{ Amps}$$

Thus, $I_1 = 5A$, $I_2 = 7A$, $I_3 = 2A$, $I_4 = 6A$, and $I_5 = 4A$.

We can confirm that Kirchoff's current law holds true around the circuit by using node C as our reference point to calculate the currents entering and leaving the junction as:

At node C $\Sigma_{IN} = \Sigma_{OUT}$

$$I_T = I_1 + I_2 = I_3 + I_4 + I_5$$

$$\therefore\ 12 = (5+7) = (2+6+4)$$

So we can confirm by analysis that Kirchhoff's current law (KCL) which states that the algebraic sum of the currents at a junction point in a circuit network is always zero is true and correct in this example.

2. Find the currents flowing around the following circuit using Kirchhoff's Current Law only.

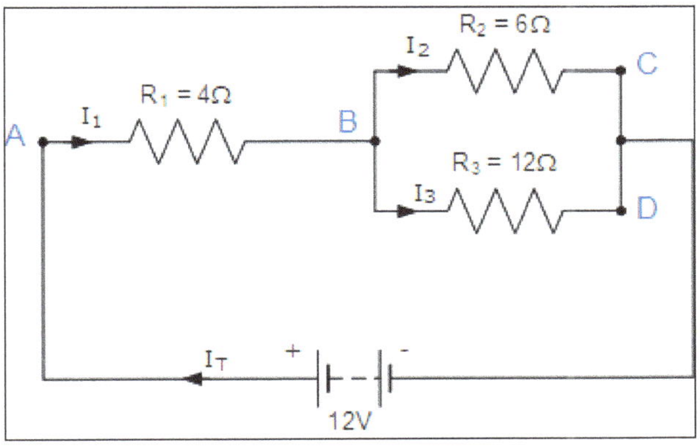

I_T is the total current flowing around the circuit driven by the 12V supply voltage. At point A, I_1 is equal to I_T, thus there will be an I_1*R voltage drop across resistor R_1.

The circuit has 2 branches, 3 nodes (B, C and D) and 2 independent loops, thus the I*R voltage drops around the two loops will be:

1. Loop ABC $\Rightarrow 12 = 4I_1 + 6I_2$
2. Loop ABD $\Rightarrow 12 = 4I_1 + 12I_3$

Since Kirchhoff's current law states that at node B, $I_1 = I_2 + I_3$, we can therefore substitute current I_1 for $(I_2 + I_3)$ in both of the following loop equations and then simplify.

Kirchhoff's Loop Equations:

Loop (ABC)	Loop (ABD)
$I_2 = 4I_1 + 6I_2$	$I_2 = 4I_1 + 12I_3$
$I_2 = 4(I_2 + I_3) + 6I_2$	$I_2 = 4(I_2 + I_3) + 12I_3$
$I_2 = 4I_2 + 4I_3 + 6I_2$	$I_2 = 4I_2 + 4I_3 + 12I_3$
$I_2 = 10I_2 + 4I_3$	$I_2 = 4I_2 + 16I_3$

We now have two simultaneous equations that relate to the currents flowing around the circuit.

$$12 = 10I_2 + 4I_3$$

$$12 = 4I_2 + 16I_3$$

By multiplying the first equation (Loop ABC) by 4 and subtracting Loop ABD from Loop ABC, we can be reduced both equations to give us the values of I_2 and I_3.

$$12 = 10I_2 + 4I_3 \ (x4) \Rightarrow 48 = 40I_2 + 16I_3$$

$$12 = 4I_2 + 16I_3 \ (x1) \Rightarrow 12 = 4I_2 + 16I_3$$

Subtracting $12 = 10I_2 + 4I_3$ and $12 = 4I_2 + 16I_3$

$$\Rightarrow 36 = 36I_2 + 0$$

Substitution of I_2 in terms of I_3 gives us the value of I_2 as 1.0 Amps,

Now we can do the same procedure to find the value of I_3 by multiplying the first equation (Loop ABC) by 4 and the second equation (Loop ABD) by 10. Again by subtracting Loop ABC from Loop ABD, we can be reduced both equations to give us the values of I_2 and I_3.

$$12 = 10I_2 + 4I_3 \ (x4) \Rightarrow 48 = 40I_2 + 16I_3$$

$$12 = 4I_2 + 16I_3 \ (x10) \Rightarrow 120 = 40I_2 + 160I_3$$

Solving above equations $\Rightarrow 72 = 0 + 144I_3$

Thus substitution of I_3 in terms of I_2 gives us the value of I_3 as 0.5 Amps,

As Kirchhoff's junction rule states that: $I_1 = I_2 + I_3$

The supply current flowing through resistor R_1 is given as: $1.0 + 0.5 = 1.5$ Amps,

Thus $I_1 = I_T = 1.5$ Amps, $I_2 = 1.0$ Amps and $I_3 = 0.5$ Amps and from that information we could calculate the I*R voltage drops across the devices and at the various points (nodes) around the circuit.

Thévenin's Theorem

Thévenin's theorem is an analytical method used to change a complex circuit into a simple equivalent circuit consisting of a single resistance in series with a source voltage.

Thévenin's Theorem states that "Any linear circuit containing several voltages and resistances can be replaced by just one single voltage in series with a single resistance connected across the load". In other words, it is possible to simplify any electrical circuit, no matter how complex, to an equivalent two-terminal circuit with just a single constant voltage source in series with a resistance (or impedance) connected to a load as shown below.

Thévenin's Theorem is especially useful in the circuit analysis of power or battery systems and other interconnected resistive circuits where it will have an effect on the adjoining part of the circuit.

Thévenin's Equivalent Circuit

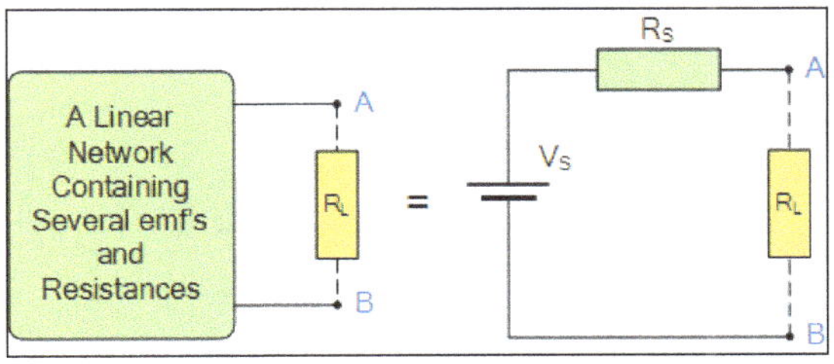

As far as the load resistor R_L is concerned, any complex "one-port" network consisting of multiple resistive circuit elements and energy sources can be replaced by one single equivalent resistance R_s and one single equivalent voltage V_s. R_s is the source resistance value looking back into the circuit and V_s is the open circuit voltage at the terminals.

For example, consider the circuit:

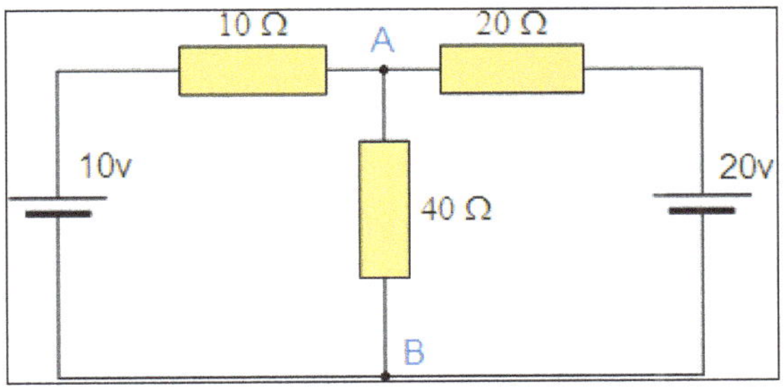

Firstly, to analyse the circuit we have to remove the centre 40Ω load resistor connected across the terminals A-B, and remove any internal resistance associated with the voltage source(s). This is done by shorting out all the voltage sources connected to the circuit, that is v = 0, or open circuit any connected current sources making i = 0. The reason for this is that we want to have an ideal voltage source or an ideal current source for the circuit analysis.

The value of the equivalent resistance, R_s is found by calculating the total resistance looking back from the terminals A and B with all the voltage sources shorted. We then get the following circuit.

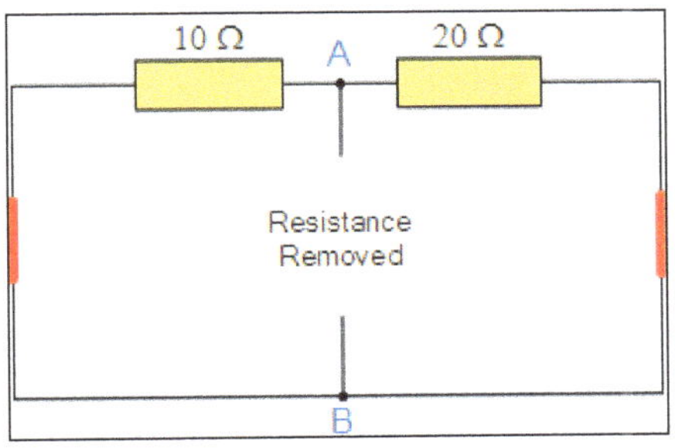

Find the Equivalent Resistance (R_s)

10Ω Resistors in parallel with the 20Ω Resistor

$$R_T = \frac{R_1 \times R_2}{R_1 \times R_2} = \frac{20 \times 10}{20 + 10} = 6.67\Omega$$

The voltage V_s is defined as the total voltage across the terminals A and B when there is an open circuit between them. That is without the load resistor R_L connected.

Find the Equivalent Voltage (V_s)

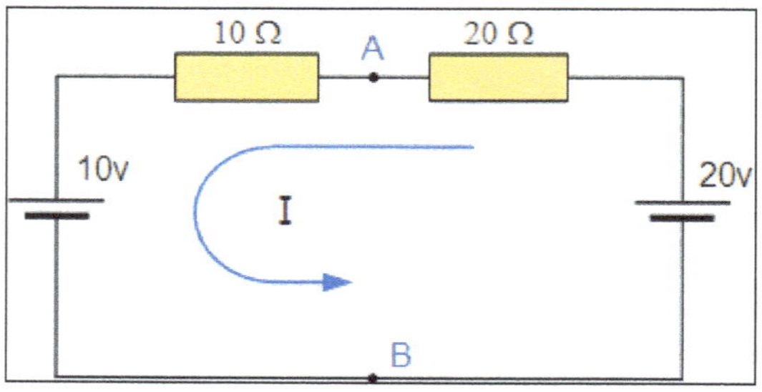

We now need to reconnect the two voltages back into the circuit, and as $V_S = V_{AB}$ the current flowing around the loop is calculated as:

$$I=\frac{V}{R}=\frac{20\text{v}-10v}{20\Omega+10\Omega}=0.33\,amps$$

This current of 0.33 amperes (330mA) is common to both resistors so the voltage drop across the 20Ω resistor or the 10Ω resistor can be calculated as:

V_{AB} = 20 – (20Ω x 0.33amps) = 13.33 volts.

or

V_{AB} = 10 + (10Ω x 0.33amps) = 13.33 volts, the same.

Then the Thevenin's Equivalent circuit would consist or a series resistance of 6.67Ω and a voltage source of 13.33v. With the 40Ω resistor connected back into the circuit we get:

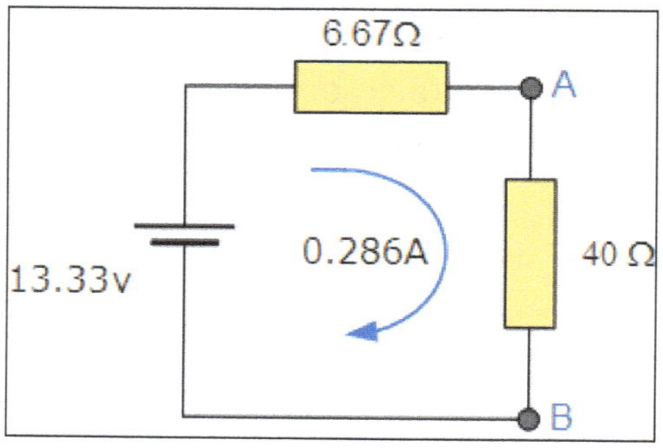

And from this the current flowing around the circuit is given as:

$$I=\frac{V}{R}=\frac{13.33}{6.67\Omega+40\Omega}=0.286\,amp$$

Which again, is the same value of 0.286 amps; we found using Kirchhoff's circuit law.

Thévenin's theorem can be used as another type of circuit analysis method and is particularly useful in the analysis of complicated circuits consisting of one or more voltage or current source and resistors that are arranged in the usual parallel and series connections.

While Thévenin's circuit theorem can be described mathematically in terms of current and voltage, it is not as powerful as Mesh Current Analysis or Nodal Voltage Analysis in larger networks because the use of Mesh or Nodal analysis is usually necessary in any The venin exercise, so it might as well be used from the start. However, Thévenin's equivalent circuits of Transistors, Voltage Sources such as batteries etc. are very useful in circuit design.

Norton's Theorem

Norton's theorem states that any two terminal linear network that constitute independent sources and linear resistances can be replaced with an equivalent circuit consisting of a current source with a parallel resistor.

Magnitude of this equivalent current source is equal to the short circuit current flowing through the load terminals and the equivalent resistance is the resistance at the load terminals, when all the sources in a given circuit are replaced by their internal resistances.

In figure a part of a network, constituting of sources (either voltage or current or both) and resistances is replaced with a current source and a parallel resistor such that current flowing through the load is same in both cases.

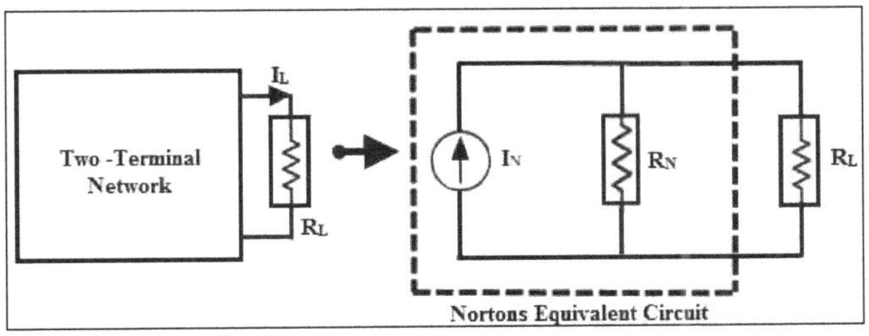

For an AC circuit it can be stated as, any active two terminal network consisting of independent sources and impedance can be replaced with an equivalent circuit consisting of a constant current source with a parallel impedance.

The value of the current source is equal to the current flow through the short circuited terminals of the network. And the parallel impedance is the equivalent impedance viewed from the short circuited terminals when all the sources are replaced with their internal impedance.

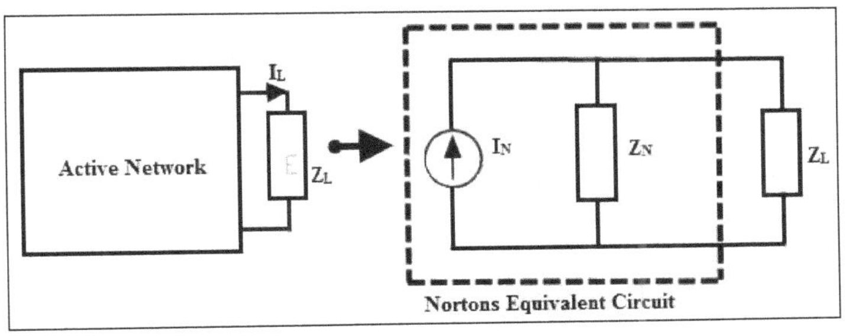

Steps to Analyse the Norton's Theorem

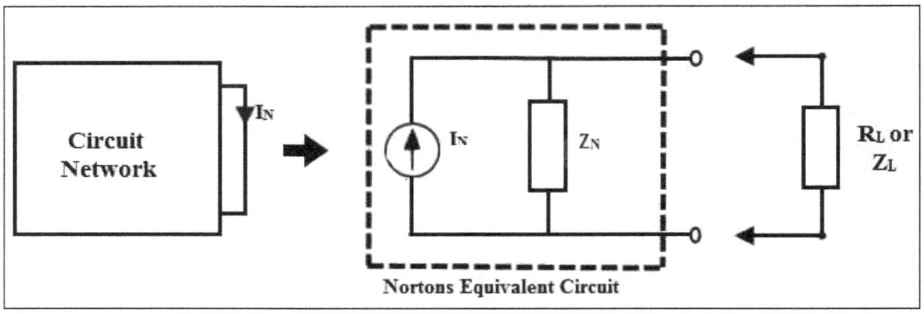

To find the load variables using Norton's theorem, Norton's equivalent parameters have to be determined. Those are Norton current or magnitude of equivalent current source and Norton resistance RN or impedance ZN. The following steps are required to determine them:

1. Consider the given circuit and short the load terminals after disconnecting the load resistance (or impedance in case of AC circuit) from output or load terminals.
2. Determine the short circuit current IN, through the shorted terminals by applying any of the circuit reduction techniques like mesh analysis or nodal analysis or superposition theorem. Or simply measure the load current using the ammeter experimentally.
3. Redraw the given circuit by replacing all the practical sources in the circuit with their internal voltages or simply short circuit voltages sources and open circuit the current sources. And also make sure to open or remove the short circuited terminals of the load.
4. Calculate the resistance (or impedance) that exists between the load terminals by looking from the load terminals. This resistance is equivalent Norton's resistance RN or (impedance ZN).
5. Insert the resistance (or impedance) in parallel with a current source IN which forms a Norton's equivalent circuit.
6. Now reconnect the load to the Norton's equivalent circuit and calculate the current, voltage and power associated with the load as.

In DC circuit:

Load current= $I_L = I_N \times [R_N / (R_L + R_N)]$

Load Voltage= $V_L = I_L \times R_L$

Power dissipated at the load= $P = I_{L2} \times R_L$

In AC circuit:

Load current= $I_L = I_N \times [Z_N / (Z_L + Z_N)]$

Load Voltage= $V_L = I_L \times Z_L$

Power dissipated at the load= $P = I_{L2} \times Z_L$

Example for Finding Equivalent Circuit to DC Circuit

Let us consider the same DC circuit in Thévenin's theorem example to apply the Norton's theorem to find the current flow through branch ab i.e., through the load resistance $R_L = R_2$-2 ohms.

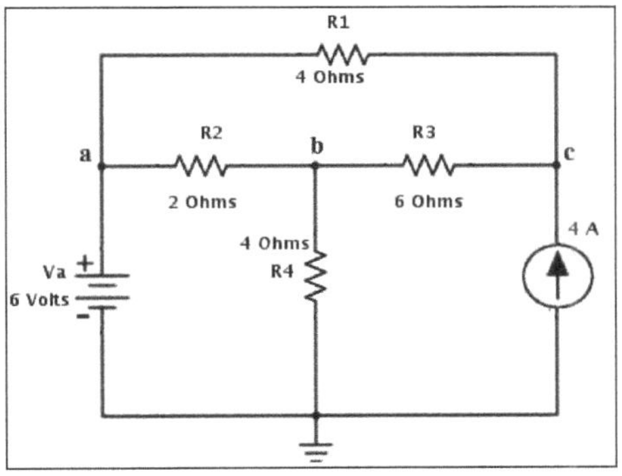

1. Disconnect the load resistance and short the load terminals a and b. Represent the current flow direction in every loop as shown in figure.

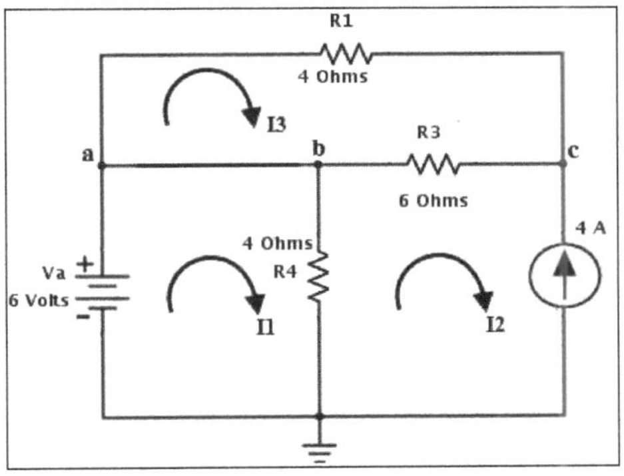

2. Apply the mesh analysis for each loop to find the current flow IN through the shorted terminals.

By applying KVL to loop 1 we get:

$6 - (I_1 - I_2)\, R_4 = 0$

Substituting:

$I_2 = -4A$

$I_1 = 6 - 16 / 4 = -2.5\ A$

By applying KVL to Loop 3 we get:

$-I_3R_1 - (I_3 - I_2)R_3 = 0$

$-4I_3 - 6\,(I_3 + 4) = 0$

$-10I_3 = 24$

$I_3 = -2.4\ A$

Therefore $I_n = I_1 - I_3$

$= -2.5 + 2.4$

$= 0.1$A which is flowing from a to b.

3. Next step is to determine the equivalent resistance RN. To compute this resistance all sources have to be replaced with their internal resistances by removing the short terminals of the load.

Then the total resistance across the terminals a and b,

$R_N = 10 \times 4 / 10 + 4$

$= 2.85$ Ohms

4. By placing above calculated current In in parallel with resistance Rn forms a Norton's equivalent circuit as shown in figure. To determine the load variables we reconnect the load resistance across the load terminals.

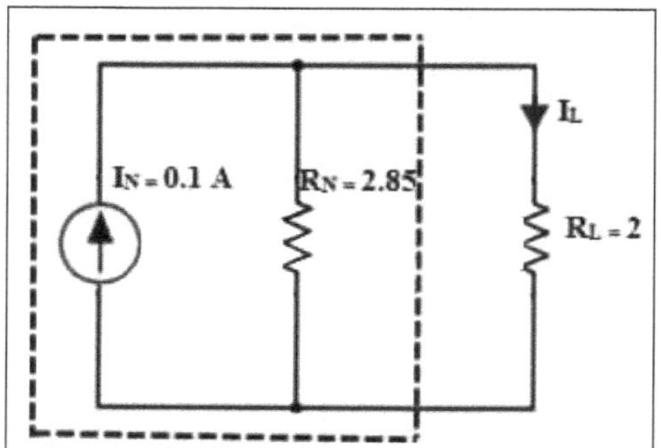

Nortons Equivalent Circuit

Then Load current= $I_L = I_N \times [R_N / (R_L + R_N)]$

$= 0.1 \times [2.85/ (2 + 2.85)]$

$= 0.05$ Amps

With the above calculated values, the original circuit is similar to the below shown figure with the representation of load current at branch ab.

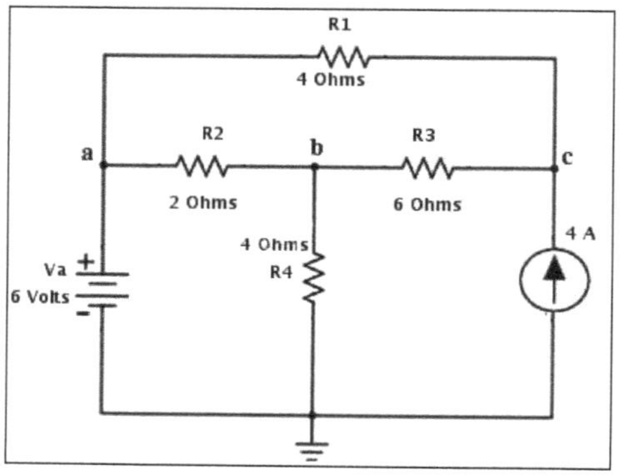

For different values of the load resistance the current flow is determined as:

When,

$R_L = 8$ ohms

$I_L = 0.1 \times [2.85 / (8 + 2.85)]$

$= 0.02$ A

When,

$R_L = 12$ ohms

$I_L = 0.1 \times [2.85 / (12 + 2.85)]$

$= 0.01$ A.

Relation between Norton's and Thevinin's Theorems

By comparing the above example with example of Thévenin's example problem, we can observe that Norton's equivalent circuit of a linear network constitute a Norton current source IN in parallel with a Thévenin's resistance R^{th}.

Therefore it is possible to perform a source transformation of Thévenin's equivalent circuit to get the Norton's equivalent circuit or vice-versa.

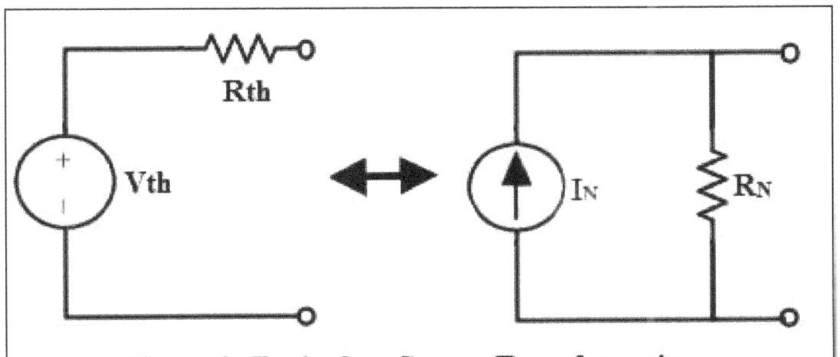

Norton's Equivalent Source Transformation

The magnitude of the voltage source (V^{th}) and a series resistance (R^{th}) from Norton's equivalent circuit using source transformation are determined as:

$V^{th} = R_N \times I_N$ and

$R^{th} = R_N$

For above example:

$V^{th} = 2.85 \times 0.1$

$= 0.28$ Volts.

Therefore, we can use any of these two methods to analyse the circuit in a simple way. However the advantages of Thévenin's theorem are also applicable to Norton's theorem as well. By using these methods one can find the current and voltage values of different load resistance values without doing any complex calculations again and again.

Hence, the Norton's theorem aids the designing much easier based on the application. The use of these two theorems is decided by the application where these equivalents are required, such as current follower circuits (use Norton's equivalent) and voltage amplifiers (Thévenin's equivalent).

Example for Finding Equivalent Circuit to AC Circuit

Consider the below AC circuit which was already analyzed using Thévenin's theorem. In this circuit we are going to find the current through the impedance 4^{+} 4_{j} ohm using Norton's theorem.

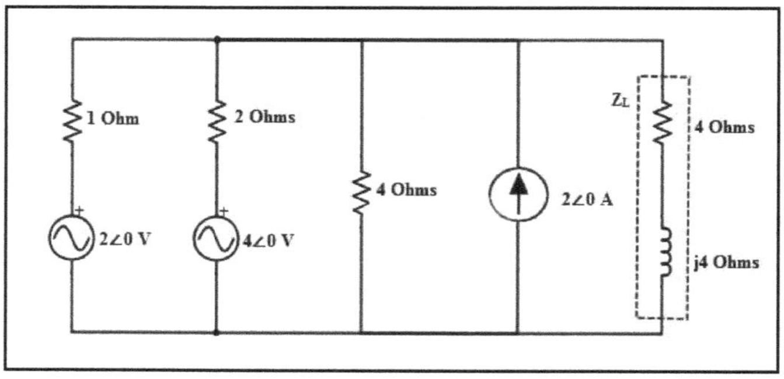

The above circuit consists of two voltage sources which can be transformed into the current source as:

$I_{s1} = V_{s1} / R_{s1}$

$= 2\angle 0 / 1$

$= 2$ A

Similarly,

$I_{s2} = V_{s2} / R_{s2}$

$= 4\angle 0 / 2$

$= 2$ A

Then the circuit becomes.

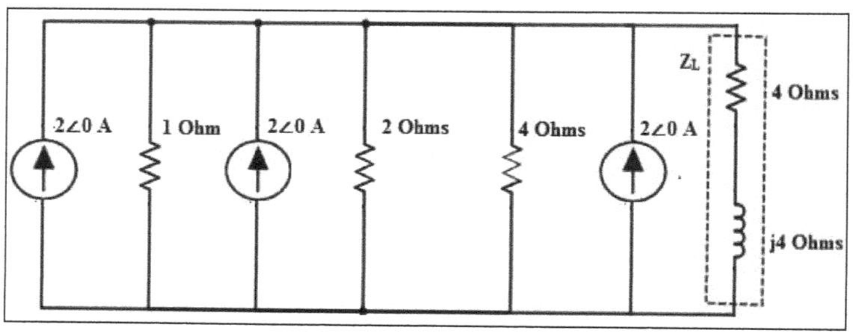

For applying Norton's theorem, we disconnect the load impedance and short the load terminals as shown in figure. Assume the current directions as represented in figure.

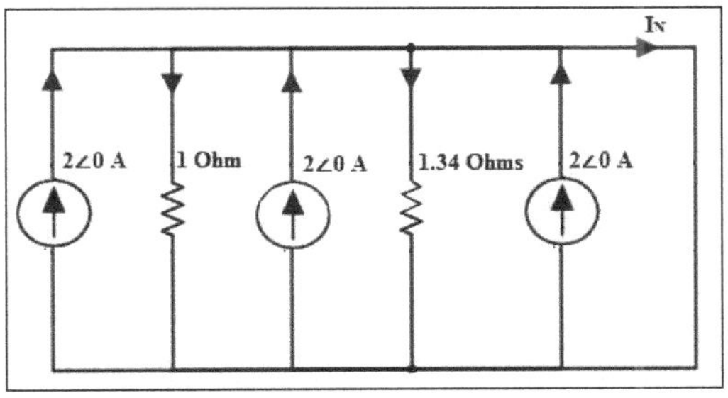

Consider the above figure as single node and the total current becomes 6 amperes and the total parallel combination of resistance is 0.574 ohms. This can be transformed into a voltage source for ease of finding the Norton's current is given as.

$V_s = 6 \angle 0 \times 0.574$

$= 3.44\angle 0$

Therefore,

$I_N = V_N / 0.574$

$= 3.44\angle 0 / 0.574$

$= 5.97\angle 0$ A

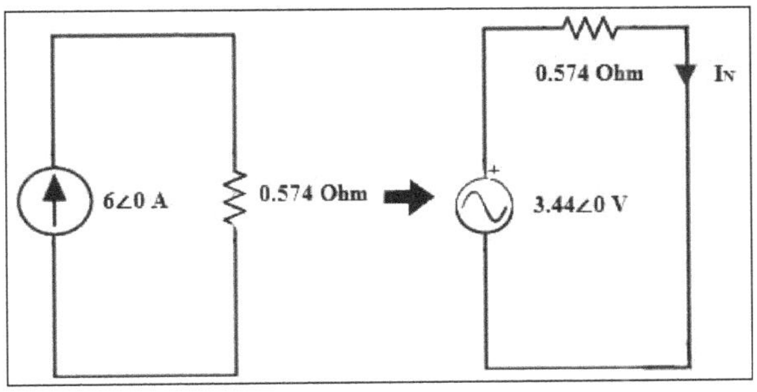

The Norton's equivalent impedance is equal to the circuit equivalent impedance, $Z_N = 0.574$.

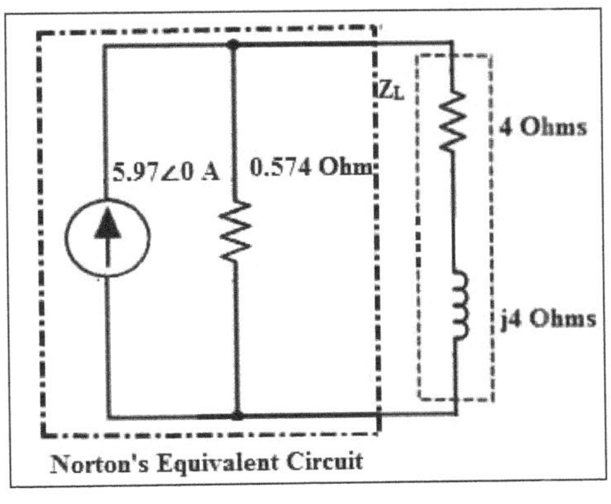

Therefore the Load current across the 4 + j4 impedance is:

$I_L = I_N \times [Z_N / (Z_L + Z_N)]$

$= 5.97\angle 0 \times [0.574 / (4 + j_4 + 0.574)]$

$= 3.42 / 6.07\angle 41.17$

$= 0.56 \angle -41.17$ A

This value is identical to the value obtained in the case Thévenin's example of AC circuit. And hence Norton's theorem is the dual of Thévenin's theorem. Limitations of Thévenin's theorem are also applicable for norton's theorem.

Superposition Theorem

The superposition theorem states that in any linear bilateral network that consisting of two or more independent sources, current through (or voltage across) an element is the algebraic sum of the currents through (voltages across) that element caused by each independent source acting alone with all other sources are replaced by their internal resistances. We know that as long as the linearity exists between the source and contribution, the total contribution due to various sources acting simultaneously is equal to the algebraic sum of individual contributions due to individual source acting at a time.

Therefore, if the circuit consists of N independent sources, we have to analyse N circuits, each will produce a result with respect to each individual source. And finally these individual results must be added to get the whole analysis of the circuit. Therefore, this require more work however, this theorem will be very useful in analysing the various parts of a complex circuit.

Steps to Analyse Superposition Theorem

- Consider the various independent sources in a given circuit.
- Select and retain one of the independent sources and replace all other sources with their internal resistances or else replace the current sources with open circuits and voltage sources with short circuits.
- To avoid confusion re-label the voltage and current notations suitably.
- Find out the desired voltage/currents due to the one source acting alone using various circuit reduction techniques.
- Repeat the steps 2 to 4 for each independent source in the given circuit.
- Algebraically add all the voltages/currents that are obtained from each individual source (Consider the voltage signs and current directions while adding).

Example:

- Let us consider the below simple DC circuit to apply the superposition theorem such that we will obtain the voltage across the resistance 10 Ohms (load terminals). Consider that in a given circuit there are two independent sources as voltage and current sources as shown in figure.

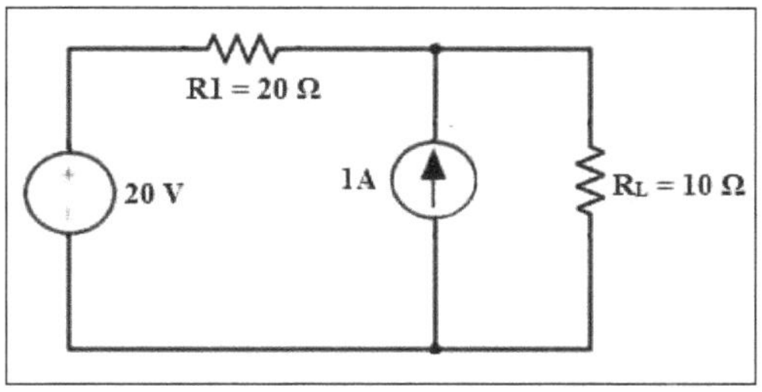

- First, we retain one source at a time that means, only voltage source is acting in the circuit and the current source is replaced with internal resistance (infinite) so it becomes open circuited as shown in figure.

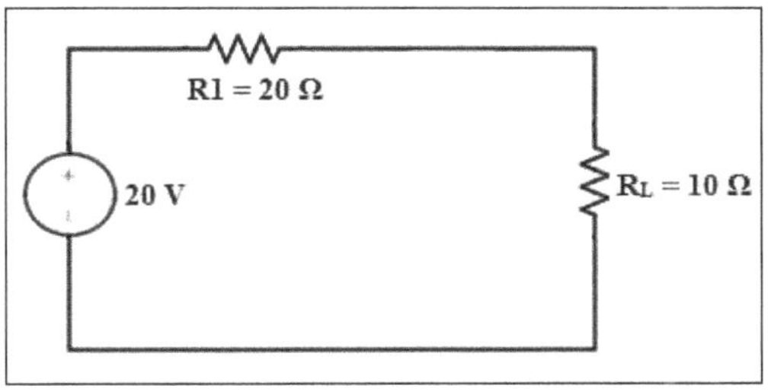

Consider V_{L1} is the voltage across the load terminals with voltage source acting alone then,

$V_{L1} = V_s \times R_L / (R_L + R_1)$

$= 20 \times 10 / (10 + 20)$

= 6. 66 Volts

- Retain the current source alone and replace the voltage source with its internal resistance (zero) so it becomes a short circuited as shown in figure.

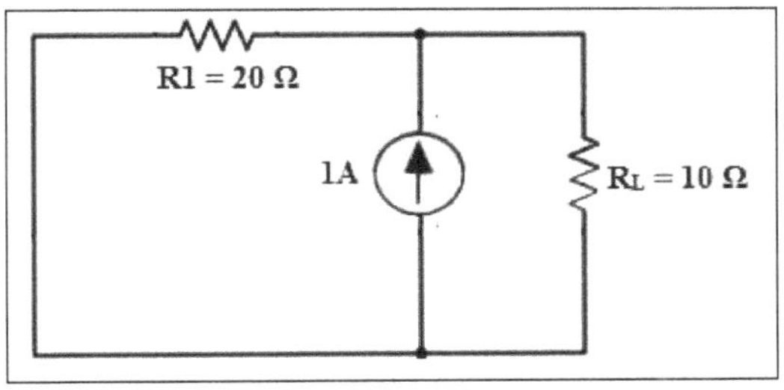

Consider that V_{L2} is the voltage across the load terminals when current source acting alone. Then,

$V_{L2} = I_L \times R_L$

$I_L = I \times R_1 / (R_1 + R_L)$

$= 1 \times 20 / (20 + 30)$

= 0.4 Amps

$V_{L2} = 0.4 \times 10$

$= 4$ Volts.

Therefore, according the superposition theorem, the voltage across the load is the sum of V_{L1} and V_{L2}:

$V_L = V_{L1} + V_{L2}$

$= 6.66 + 4$

$= 10.66$ Volts.

Example:

Consider the below circuit to which we are going to determine the current I through the 4 ohm resistor using superposition theorem.

Consider I1, I2 and I3 are the currents due to sources 12v, 20V and 4A sources respectively. Then, based superposition theorem I = I1 + I2 + I3. So let's determine these currents with each individual source.

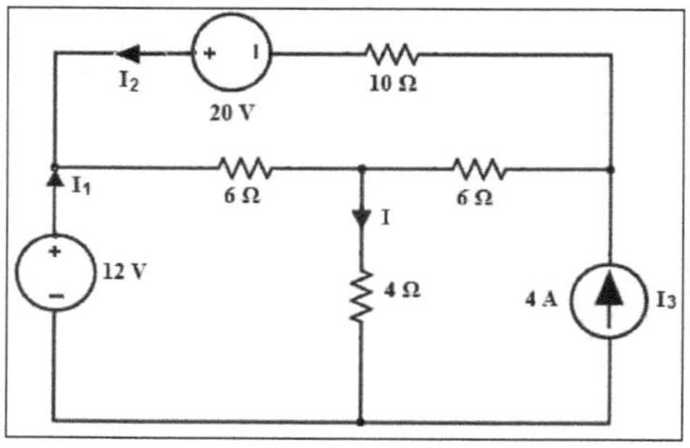

Only with 12V Voltage source

Consider the below circuit where only 12V source is retained in the circuit and other sources are replaced by their internal resistances.

By combining the resistance 6 ohm with 10 ohm we get 16 ohm resistance which is parallel with 6 ohm resistance. Then this combination produce, 16 × 6 / (16 + 6) = 4.36 ohm. Therefore the equivalent circuit will be as shown in figure.

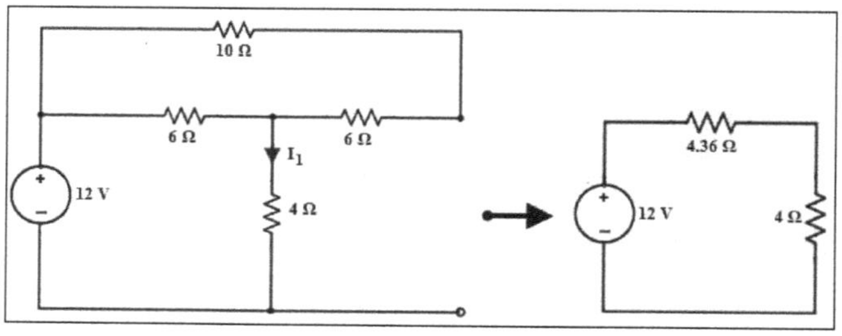

Then the current through 4 ohms resistance,

$I_1 = 12 / 8.36$

$= 1.43$ A

Only with 20V Voltage Source

Retain only 20V voltage source and replace other sources with their internal resistance, then the circuit becomes as shown below.

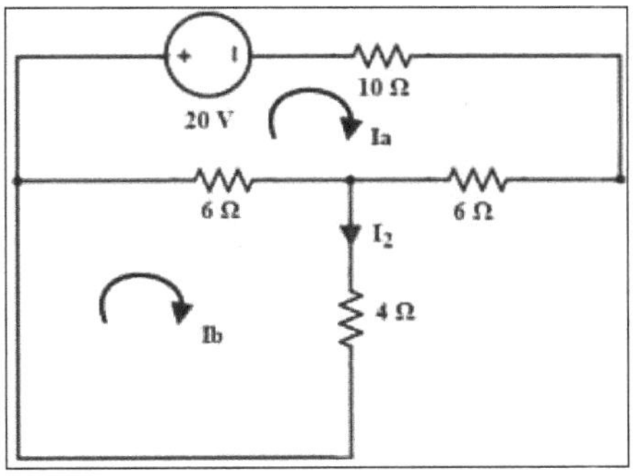

Apply the mesh analysis to the loop a, we get:

$$22I_a - 6I_b + 20 = 0$$

$$22I_a - 6I_b = -20$$

For Loop b, we get:

$$10I_b - 6I_a = 0$$

$$Ia = 10I_b/6$$

Substituting Ib in equation ($22Ia - 6I_b = -20$):

$$22\,(10I_b/6) - 6I_b = -20$$

$$I_b = -\,0.65$$

Therefore,

$$I_2 = I_b = -0.65.$$

Only with 4A Current Source

Consider the below circuit where only current source is retained and other sources are replaced with their internal resistances.

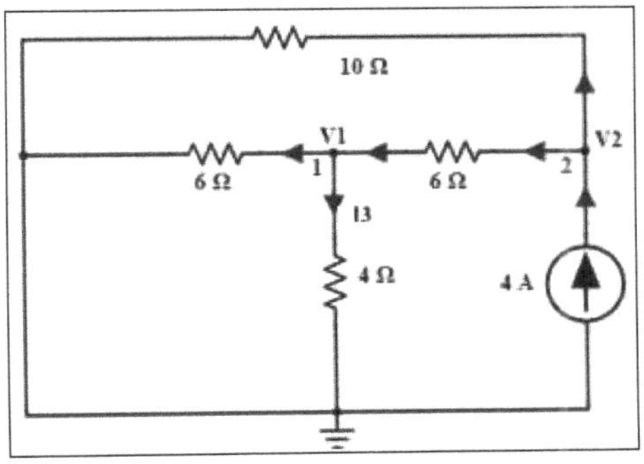

By applying nodal analysis at node 2 we get:

$$4 = (V_2/10) + (V_2 - V_1)/6$$

At node1:

$$(V_1/6) + (V_1/4) = (V_2 - V_1) / 6$$

$$V_2 = 3.496\ V_1$$

Substituting V_2 in equation $(4 = (V_2/10) + (V_2 - V_1)/6)$, we get:

$$V_1 = 0.766 \text{ Volts.}$$

Therefore,

$$I_3 = V_1/4$$

$$= 0.766/4$$

$$= 0.19 \text{ Amps.}$$

Thus, as per the superposition theorem:

$$I = I_1 + I_2 + I_3$$

$$= 1.43 - 0.65 + 0.19$$

$$= 0.97 \text{ Amps.}$$

Superposition Example using AC Circuit

Consider the below AC circuit, to which we are going to determine the value of current in the 4 ohm resistor using superposition theorem.

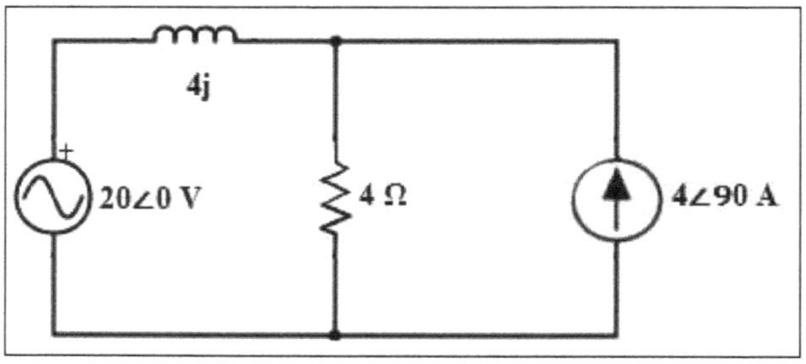

Case 1: Only with 20∠0 Voltage Source

By retaining the voltage source alone in the circuit, the current flow through the circuit is determined as:

$$I_1 = 20\angle 0 / (4 + j_4)$$

$$= 20\angle 0 / (5.65\angle 45)$$

$$= 3.53\angle{-45} \text{ or } 2.49 - j2.49 \text{ A}$$

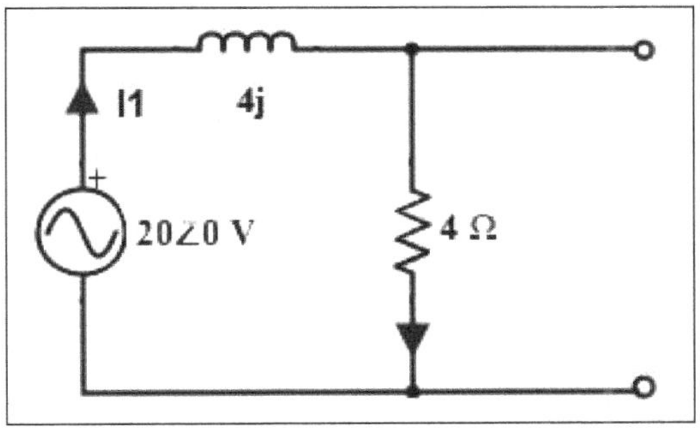

Case 2: Only with 4∠90 Current Source

By retaining the current source alone in the circuit, the current I2 through the circuit is determined as:

By current division method, $I_2 = 4\angle 90 \times 4j / (4 + j4)$

$= 4\angle 90 \times 4\angle 90\ (5.65\angle 45)$

$= 4\angle 90 \times 0.707\angle 45$

$= 2.828\angle 135$ or $-1.99 + j1.99$ A

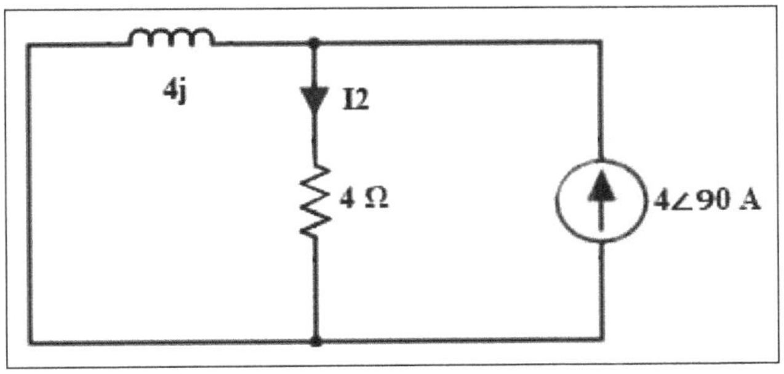

The resultant current through the resistor 4 ohms is:

$I = I_1 + I_2$

$= 3.53\angle -45 + 2.828\angle 135$

$= 0.785\angle 45$ or $0.56 + j0.56$ A.

Limitations of Superposition Theorem

1. For power calculations superposition theorem cannot be used as this theorem works based on the linearity. Because the power equation is not linear as it is the product of voltage and current or square of the current or square of the voltage. Thus the power consumed by the element in a given circuit with superposition theorem is not possible.

2. If the choice of the load is variable or the load resistance changes frequently, then it is required to perform every source contribution of current or voltage and their sum for every change in load resistance. So this very complex procedure for analysing complex circuits.

3. This theorem applicable for only linear circuits and for non-linear circuits (Having transistors and diodes) we cannot apply.
4. This theorem is applicable only if the circuit has more than one source.

Reciprocity Theorem

Reciprocity Theorem states that – In any branch of a network or circuit, the current due to a single source of voltage (V) in the network is equal to the current through that branch in which the source was originally placed when the source is again put in the branch in which the current was originally obtained. This theorem is used in the bilateral linear network which consists of bilateral components.

In simple words, we can state the reciprocity theorem as when the places of voltage and current source in any network are interchanged the amount or magnitude of current and voltage flowing in the circuit remains the same. This theorem is used for solving many DC and AC network which have many applications in electromagnetism electronics. Their circuit does not have any time varying element.

The location of the voltage source and the current source may be interchanged without a change in current. However, the polarity of the voltage source should be identical with the direction of the branch current in each position.

The Reciprocity Theorem is explained with the help of the circuit diagram.

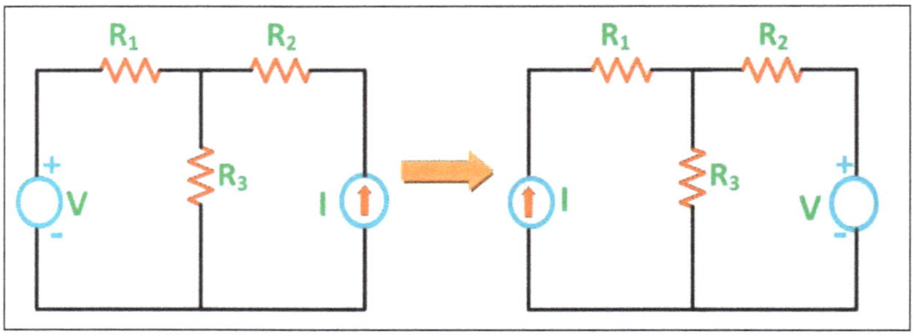

The various resistances R_1, R_2, R_3 is connected in the circuit diagram above with a voltage source (V) and a current source (I). It is clear from the figure above that the voltage source and current sources are interchanged for solving the network with the help of Reciprocity Theorem.

The limitation of this theorem is that it is applicable only to single source networks and not in the multi-source network. The network where reciprocity theorem is applied should be linear and consist of resistors, inductors, capacitors and coupled circuits. The circuit should not have any time-varying elements.

Steps for Solving a Network Utilizing Reciprocity Theorem

- Step 1 – Firstly, select the branches between which reciprocity has to be established.

- Step 2 – The current in the branch is obtained using any conventional network analysis method.
- Step 3 – The voltage source is interchanged between the branch which is selected.
- Step 4 – The current in the branch where the voltage source was existing earlier is calculated.
- Step 5 – Now, it is seen that the current obtained in the previous connection, i.e., in step 2 and the current which is calculated when the source is interchanged, i.e., in step 4 are identical to each other.

Maximum Power Transfer Theorem

In any electric circuit, the electrical energy from the supply is delivered to the load where it is converted into a useful work. Practically, the entire supplied power will not present at load due to the heating effect and other constraints in the network. Therefore, there exist a certain difference between drawing and delivering powers.

The load size always affects the amount of power transferred from the supply source, i.e., any change in the load resistance results to change in power transfer to the load. Thus, the maximum power transfer theorem ensures the condition to transfer the maximum power to the load.

Maximum Power Transfer Theorem Statement

The maximum power transfer theorem states that in a linear, bilateral DC network, maximum power is delivered to the load when the load resistance is equal to the internal resistance of a source.

If it is an independent voltage source, then its series resistance (internal resistance Rs) or if it is independent current source, then its parallel resistance (internal resistance Rs) must equal to the load resistance R_L to deliver maximum power to the load.

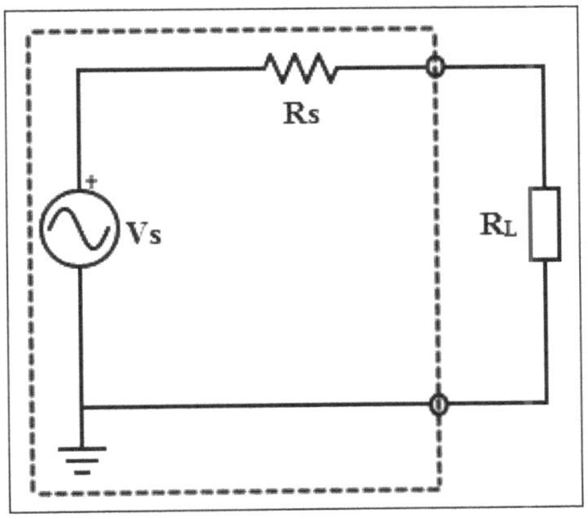

Proof of Maximum Power Transfer Theorem

The maximum power transfer theorem ensures the value of the load resistance, at which the maximum power is transferred to the load.

Consider the below DC two terminal network (left side circuit), to which the condition for maximum power is determined , by obtaining the expression of power absorbed by load with use of mesh or nodal current methods and then derivation the resulting expression with respect to load resistance R_L.

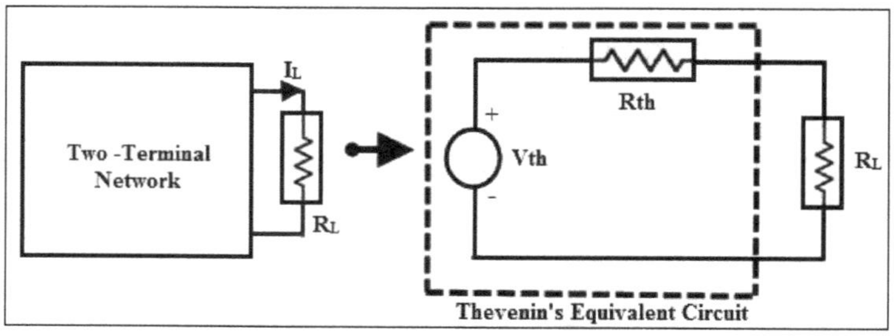

The original two terminal circuit is replaced with a Thevenin's equivalent circuit across the variable load resistance. The current through the load for any value of load resistance is:

$$I_L = \frac{V_{Th}}{R_{Th} + R_L}$$

The power absorbed by the load is:

$$P_L = I_L^2 \times R_L$$

$$= \left[\frac{V_{Th}}{R_{Th} + R_L}\right]^2 \times R_L$$

Form the above expression the power delivered depends on the values of R_{TH} and R_L. However the Thevenin's equivalent is constant, the power delivered from this equivalent source to the load entirely depends on the load resistance R_L. To find the exact value of R_L, we apply differentiation to P_L with respect to R_L and equating it to zero as:

$$\frac{dP(R_L)}{dR_L} = V_{Th}^2 \left[\frac{(R_{Th} + R_L)^2 - 2R_L \times (R_{Th} + R_L)}{(R_{Th} + R_L)^4}\right] = 0$$

$$\Rightarrow (R_{Th} + R_L) - 2R_L = 0$$

$$\Rightarrow R_L = R_{Th}$$

Therefore, this is the condition of matching the load where the maximum power transfer occurs when the load resistance is equal to the Thevenin's resistance of the circuit. By substituting the R_{th} = R_L in equation $I_L = \frac{V_{Th}}{R_{Th} + R_L}$ we get, the maximum power delivered to the load is:

$$P_{\max} = \left[\frac{V_{Th}}{R_{Th} + R_L}\right]^2 \times R_L \Bigg|_{R_L = R_{Th}}$$

$$= \frac{V_{Th}^2}{4R_{Th}}$$

Total power transferred from source is:

$$P_T = I_L^2 (R_{TH} + R_L)$$

$$= 2\, I_L^2 R_L$$

Hence, the maximum power transfer theorem expresses the state at which maximum power is delivered to the load, that is, when the load resistance is equal to the Thevenin's equivalent resistance of the circuit. Below figure shows a curve of power delivered to the load with respect to the load resistance.

The power delivered is zero when the load resistance is zero as there is no voltage drop across the load during this condition. Also, the power will be maximum, when the load resistance is equal to the internal resistance of the circuit (or Thevenin's equivalent resistance). Again, the power is zero as the load resistance reaches to infinity as there is no current flow through the load.

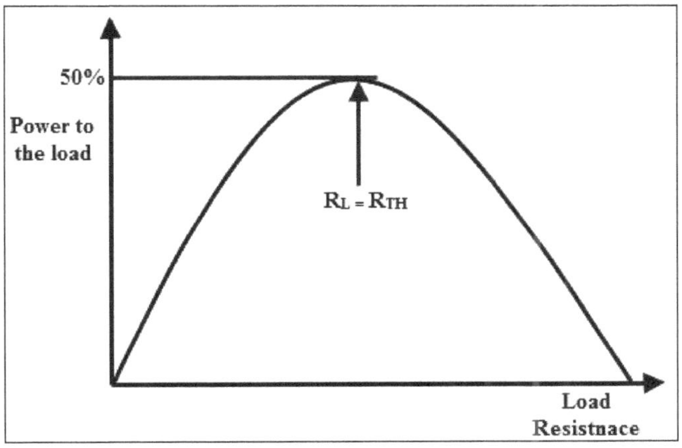

Power Transfer Efficiency

We must remember that this theorem results maximum power transfer but not a maximum efficiency. If the load resistance is smaller than source resistance, power dissipated at the load is reduced while most of the power is dissipated at the source then the efficiency becomes lower.

Consider the total power delivered from source equation ($2\, I_L^2 R_L$), in which the power is dissipated in the equivalent Thevenin's resistance R_{TH} by the voltage source V_{TH}.

Therefore, the efficiency under the condition of maximum power transfer is:

Efficiency = Output/Input × 100

$$= I_L^2 R_L / 2\, I_L^2 R_L \times 100$$

$$= 50\ \%.$$

Hence, at the condition of maximum power transfer, the efficiency is 50%, that means a half percentage of generated power is delivered to the load and at other conditions small percentage of power is delivered to the load , as indicated in efficiency verses maximum power transfer the curves in the figure.

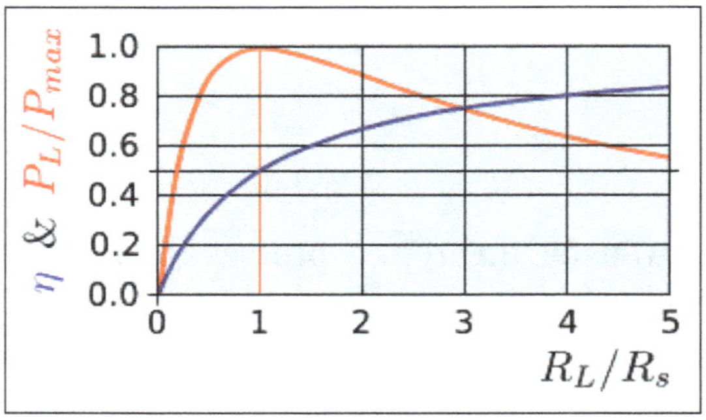

For some applications, it is desirable to transfer maximum power to the load than achieving high efficiency such as in amplifiers and communication circuits.

On the other hand, it is desirable to achieve higher efficiency than maximised power transfer in case of power transmission systems where a large load resistance (much larger value than internal source resistance) is placed across the load. Even though the efficiency is high the power delivered will be less in those cases.

Maximum Power Transfer Theorem for AC Circuits

It can be stated as in an active network, the maximum power is transferred to the load when the load impedance is equal to the complex conjugate of an equivalent impedance of a given network as viewed from the load terminals.

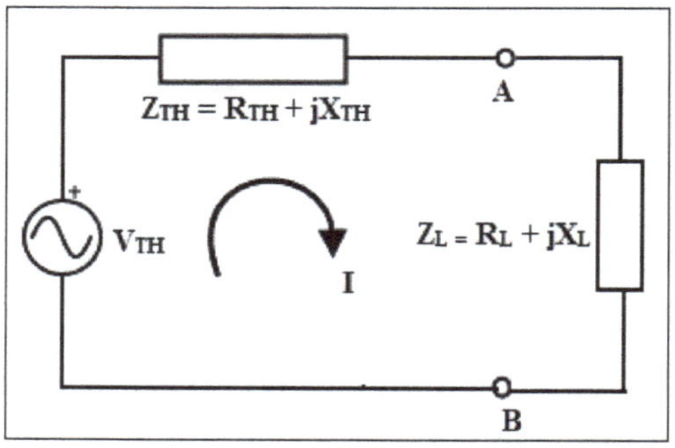

Consider the above Thevenin's equivalent circuit across the load terminals in which the current flowing through the circuit is given as:

$$I = V_{TH} / Z_{TH} + Z_L$$

Where,

$$Z_L = R_L + jX_L$$

$$Z_{TH} = R_{TH} + jX_{TH}$$

Therefore,

$$I = V_{TH}/(R_L + jX_L + R_{TH} + jX_{TH})$$

$$= V_{TH}/((R_L + R_{TH}) + j(X_L + X_{TH})).$$

The power delivered to the load:

$$P_L = I^2 R_L$$

$$P_L = V^2_{TH} \times R_L / ((R_L + R_{TH})^2 + (X_L + X_{TH})^2).$$

For maximum power the derivative of the above equation must be zero, after simplification we get:

$$X_L + X_{TH} = 0$$

$$X_L = - X_{TH}.$$

Putting the above relation in equation ($P_L = V^2_{TH} \times R_L / ((R_L + R_{TH})^2 + (X_L + X_{TH})^2)$), we get:

$$P_L = V^2_{TH} \times R_L / ((R_L + R_{TH})^2.$$

Again for maximum power transfer, derivation of above equation must be equal to zero, after simplification we get:

$$R_L + R_{TH} = 2 R_L$$

$$R_L = R_{TH}$$

Hence, the maximum power will transferred to the load from source, if $R_L = R_{TH}$ and $X_L = - X_{TH}$ in an AC circuit. This means that the load impedance should be equal to the complex conjugate of equivalent impedance of the circuit:

$$Z_L = Z_{TH}$$

Where Z_{TH} is the complex conjugate of the equivalent impedance of the circuit.

This maximum power transferred, $P_{max} = V^2_{TH} / 4 R_{TH}$ or $V^2_{TH} / 4 R_L$.

Applying Maximum Power Transfer Example to DC circuit

Consider the below circuit to which we determine the value of the load resistance that receives the maximum power from the supply source and the maximum power under the maximum power transfer condition.

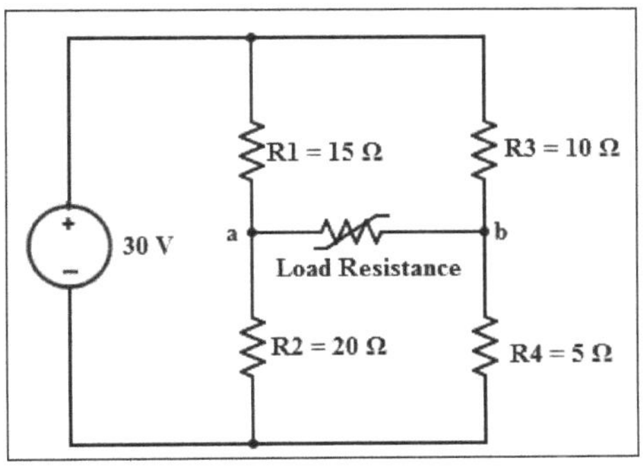

1. Disconnect the load resistance from the load terminals a and b. To represent the given circuit as Thevenin's equivalent, we are to determine the Thevenin's voltage V_{TH} and Thevenin's equivalent resistance R_{TH}.

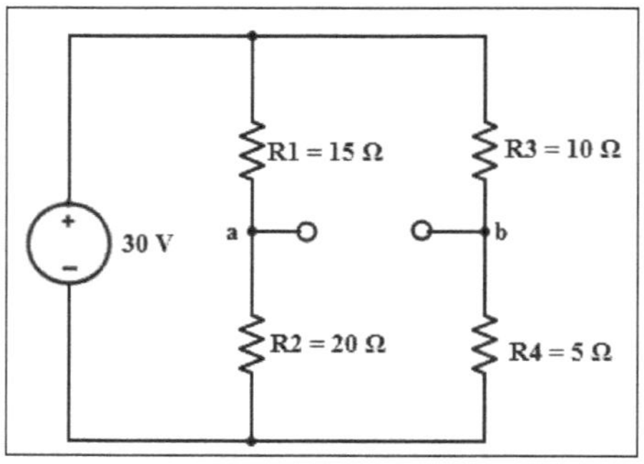

The Thevenin's voltage or voltage across the terminals ab is:

$V_{ab} = V_a - V_b$

$V_a = V \times R_2 / (R_1 + R_2)$

$= 30 \times 20 /\times (20 + 15)$

$= 17.14$ V

$V_b = V \times R_4/ (R_3 + R_4)$

$= 30 \times 5/(10 + 5)$

$= 10$ V

$V_{ab} = 17.14 - 10$

$= 7.14$ V

$V_{TH} = V_{ab} = 7.14$ Volts.

2. Calculate the Thevenin's equivalent resistance R_{TH} by replacing sources with their internal resistances (here assume that voltage source has zero internal resistance so it becomes a short circuited).

Thevenin's equivalent resistance or resistance across the terminals ab is:

$R_{TH} = R_{ab} = [R_1R_2 / (R_1 + R_2)] + [R_3R_4/(R_3 + R_4)]$

$= [(15 \times 20) / (15 + 20)] + [(10 \times 5) / (10 + 5)]$

$= 8.57 + 3.33$

$R_{TH} = 11.90$ Ohms.

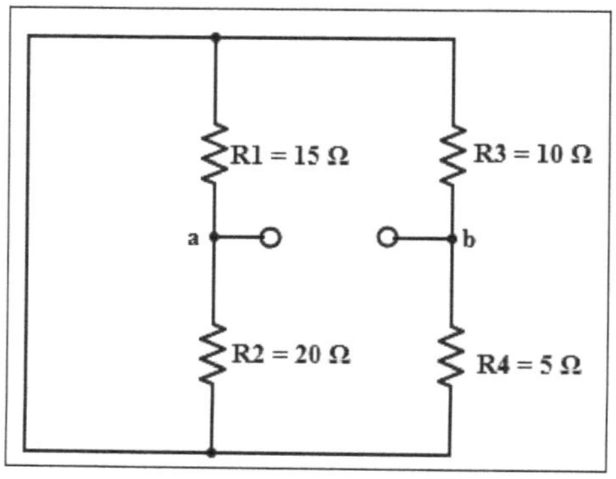

3. The Thevenin's equivalent circuit with above calculated values by reconnecting the load resistance is shown in the figure.

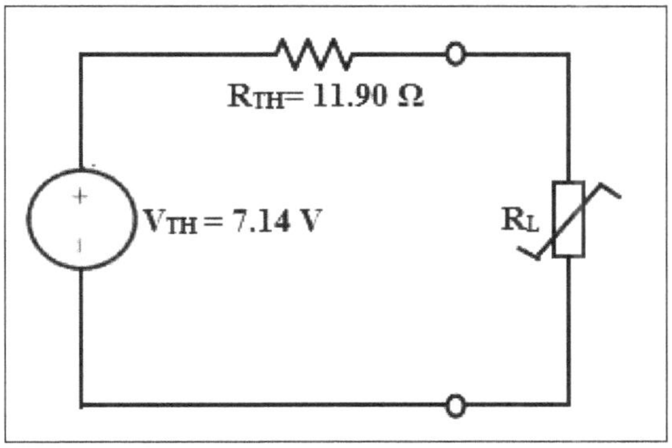

From the maximum power transfer theorem, RL value must equal to the RTH to deliver the maximum power to the load.

Therefore,

$R_L = R_{TH} = 11.90$ Ohms

And the maximum power transferred under this condition is:

$P_{max} = V^2_{TH} / 4\ R_{TH}$

$= (7.14)^2 / (4 \times 11.90)$

$= 50.97 / 47.6$

$= 1.07$ Watts.

Applying Maximum Power Transfer to AC circuit

The below AC network consists of load impedance Z_L of which both reactive and resistive parts can be varied. Hence, we have to determine the load impedance value at which the maximum power delivered from the source and the value of maximum power.

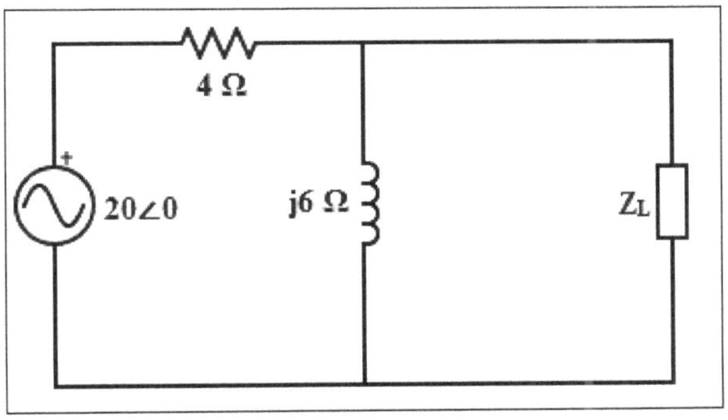

To find the value of load impedance, first, we find the Thevenin's equivalent circuit across the load terminals. For finding Thevenin's voltage, disconnect the load impedance as shown in the figure.

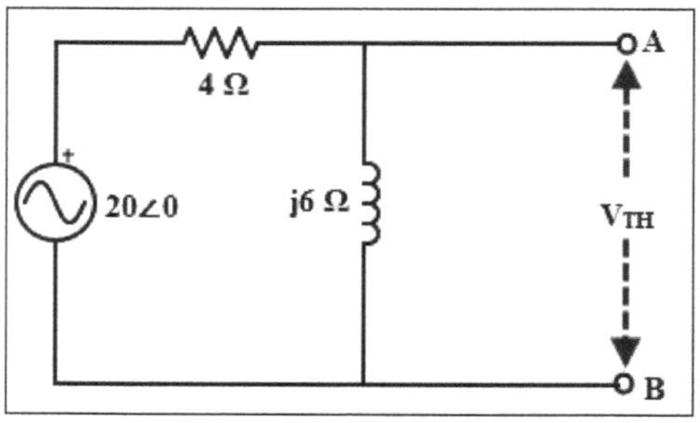

By voltage divider rule:

V_{TH} = 20∠0 × [j6 / (4 + j6)]

= 20∠0 × [6∠90 / 7.21∠56.3]

= 20∠0 × 0.825∠33.7

V_{TH} = 16.5∠33.7 V.

By shorting the voltage source, we calculate the Thevenin's equivalent impedance of the circuit as shown in figure.

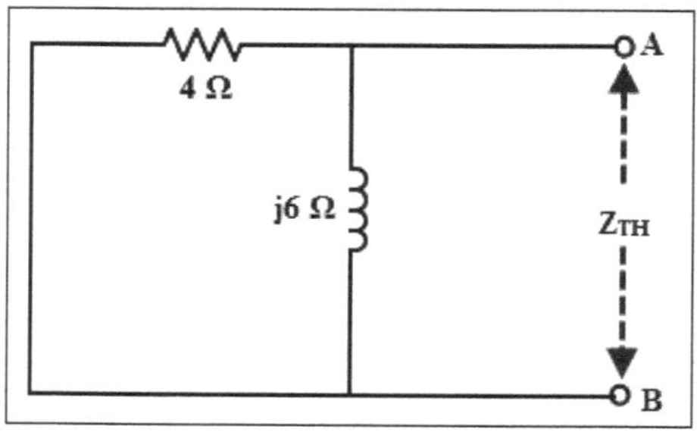

Therefore,

Z_{TH} = (4 × j6) / (4 + j6)

= (4 × 6∠90) / (7.21∠56.3)

= 3.33∠33.7 or 2.77 + j1.85 Ohms.

Hence, the Thevenin's equivalent circuit across the load terminals is shown in the figure.

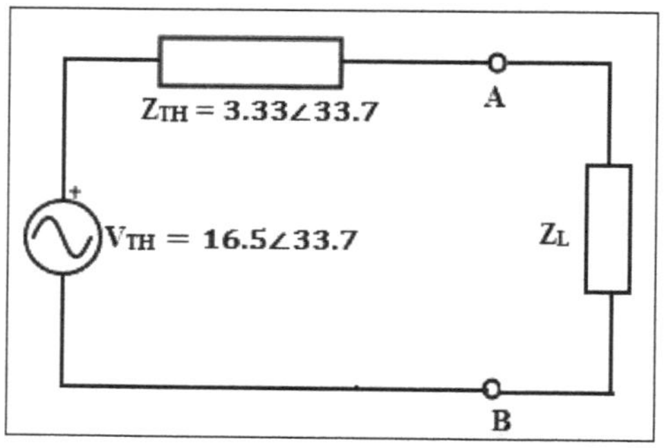

Therefore to transfer the maximum power to the load, the value of the load impedance should be:

$Z_L = Z\times_{TH}$

= 2.77 – j1.85 ohms

= 2.77 – j1.85 ohms.

The maximum power delivered:

$P_{max} = V^2_{TH} / 4 R_{TH}$

$= (16.5)^2/4(2.77)$

= 272.25 / 11.08

= 24.5 W.

Practical Application of Maximum Power Transfer Theorem

Consider the practical example of a speaker with an impedance of 8 ohms is driven by audio amplifier with its internal impedance of 500 ohms. The Thevenin's equivalent circuit is also shown in figure.

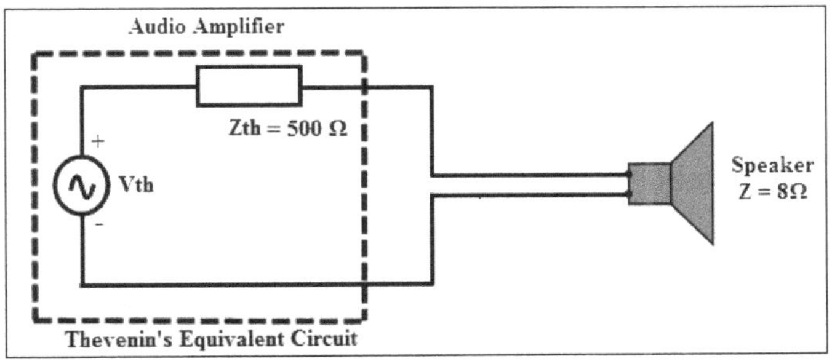

According to the maximum power transfer theorem, the power is maximized at the load if the load impedance is 500 ohms (same as internal impedance). Or else internal resistance has to be changed to 8 ohms to achieve the condition however it is not possible. So it is an impedance mismatch condition and it can be overcome by using an impedance matching transformer with its impedance transformation ratio of 500:8.

References

- Ohms-law-and-resistance, electricity, physics: toppr.com, Retrieved 2 February, 2019
- Kirchhoffs-voltage-law-kvl, direct-current: allaboutcircuits.com, Retrieved 12 June, 2019
- Kirchhoffs-current-law, dccircuits: electronics-tutorials.ws, Retrieved 18 August, 2019
- Dccircuits: electronics-tutorials.ws, Retrieved 29 May, 2019
- Nortons-theorem: electronicshub.org, Retrieved 9 March, 2019
- Superposition-theorem: electronicshub.org, Retrieved 10 January, 2019
- What-is-reciprocity-theorem: circuitglobe.com, Retrieved 16 July, 2019
- Maximum-power-transfer-theorem: electronicshub.org, Retrieved 11 April, 2019

4

Electronic Systems and Control Systems

A physical interconnection of different components which is involved in gathering various amounts of information together is known as an electronic system. A system which manages, directs and regulates the behavior of other devices is known as a control system. The chapter closely examines these key concepts of electronic and control systems to provide an extensive understanding of the subject.

Electronic Systems

Every electronic device we use in our daily life is one type of system. The systems working depending up on electronic variables like power or voltage or current is known as Electronic system. Number of blocks having different input – output relationship are connected together to form a system, which is supposed to execute a particular task or process.

Sometimes an electronic system is also referred as "Control system". A system that is used to control or regulate the output of the other system is called control system. This is also a group or collection of different systems and blocks.

The industrial automations systems which are used in heavy industries are the example of control system. These are used to control the production process in industries.

The block diagram of a basic control system, with control loop is shown below.

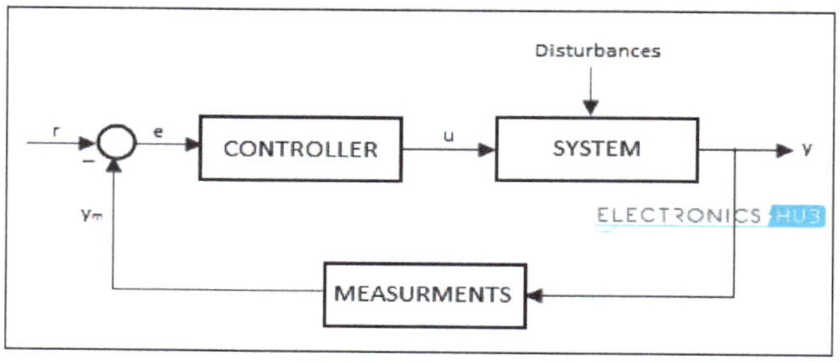

The input is connected to the controller to control the system output and the output of the controller is connected to the system or processor, which processes the output.

If the produced system output is deviated from the desired output, indicates that the system is encountered with some noise or error. The error signal is connected back to the controller, to make changes in the system according to system requirements.

A system can be represented in many ways such as pictorial manner, descriptive manner, schematic and mathematical manner etc. But many of the electronic systems are schematically represented by inter – connecting the individual blocks in series.

Each and every block of the system has its individual inputs and outputs. Each block in a control system will represent a system or an individual component of a system.

Block Diagram of System

The block diagram for the simple system is shown below, with individual inputs and outputs.

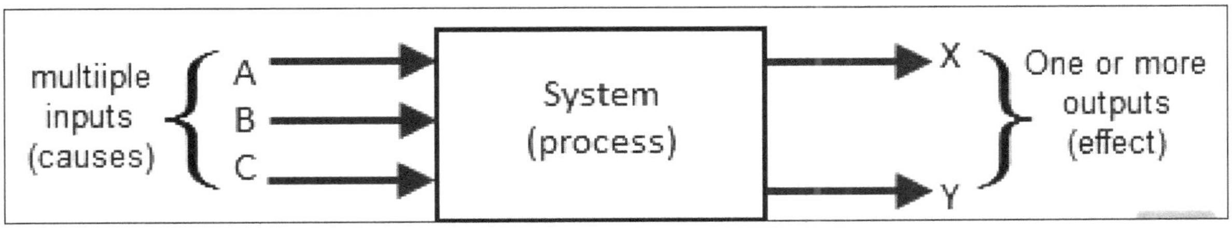

The electronic system may have multiple inputs and multiple outputs. The system accepts the inputs and it processes the required task to produce the outputs. The state of the system can be changed by processing the inputs. So the inputs are the "cause" of change in system's equilibrium state.

This means the inputs and outputs of a system are in an direct relationship that inputs will show effect on outputs. This analysis of control system is called as "Cause and Effect analysis".

An audio system, TV, or any Mp3 player is the example of an electronic system. The sound waves are converted to electrical signals and they are amplified to process the output of high strengthened audio signals, through speakers.

Example: We come across so many systems in our day to day life, such as washing machine, drier, keypad, Air conditioner etc. we use control systems mostly in industrial purposes like space technology, weapons, power systems and quality controlling of products like industrial automation etc.

Types of Electronic Systems

As we already explained earlier, every control system will have its own input and output. The input signal may be of continuous and discrete. Based on the input provided to the system, the control systems are classified into 2 types. They are:

1. Continuous system
2. Discrete system

Continuous System

Continuous system is that in which the input signals are defined as the analog signals. The analog signals are continuous in nature, so they are called as "continuous time signals".

The magnitude of the continuous signal will vary with the time period T. As the input is Analogues in nature, the continuous electronic systems produce the analogous output. This is the reason they are notified as "Pure analog systems".

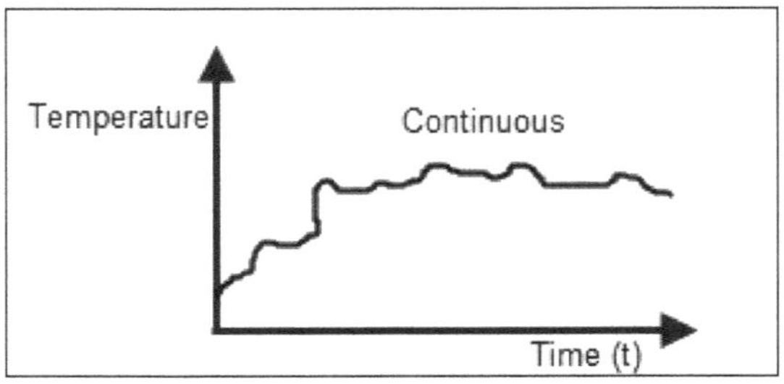

The values of the continuous electronic system varies only with time.

Example for Continuous System

The process of measuring the temperature of a room is an example of continuous system. As the temperature is measured in a range of time duration, like Monday to Saturday or temperature levels from cold to hot etc., the temperature monitoring system is analog in nature or it is said to be an example of continuous electronic system.

Other examples of the continuous electronic systems are current and voltage signals, pressure, velocity and temperature signal etc.

Discrete System

The discrete system is defined as the system having the sequence of discrete time values as its input. The values obtained at particular intervals of time are called "Discrete values". As the input of discrete system is the sequence of instinctual values, the output will be in the form of sequence of numbers.

So these discrete systems are also known as "Digital systems". Generally the output of digital system will be a binary number sequence, such as the collection of 0's and 1's.

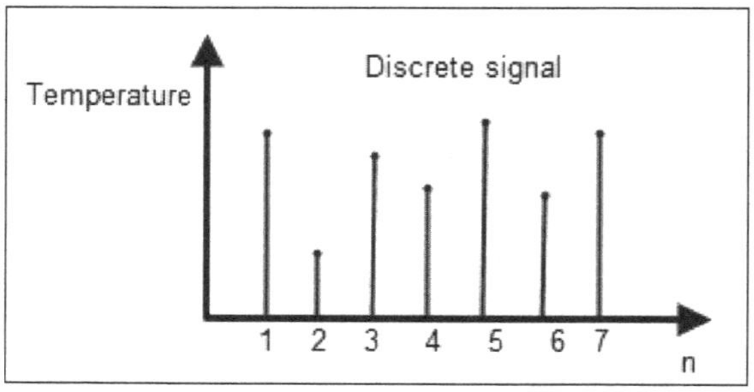

One important thing about the discrete system is that, the discrete values defined at particular time intervals must be spaced equally. For example, if the time periods are t1, t2, t3, t4 and t5, then t1 – t2 = t2 – t3 = t3 – t4 = t4 – t5. The values of discrete system don't vary over the time period, but changes at particular instances of time.

Example for Discrete system

The temperature measuring (monitoring) system which measures the temperature of room only at particular intervals of times, such as particular days like on Monday, on Tuesday or in particular time intervals like 1pm, at 2pm, at 3pm etc. is the example of discrete time electronic system.

Representing continuous system as a set of discrete intervals. We can represent the continuous electronic system into discrete electronic system, by changing the domain of the system. This means, a continuous system with input and output signals as X (t) and Y (t) can be represented as discrete electronic system with input and output as X (n) and Y (t), where X (t) and Y (t) represents the time domain and X (n) and Y (n) represents the frequency domain.

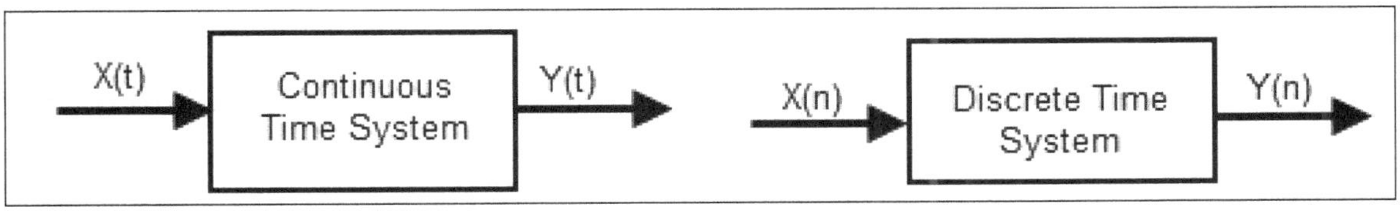

Interconnection of Systems

The blocks of the electronic system can be connected with one another, to form more complex systems. This is called "Interconnection of blocks". These interconnected blocks are about to perform the specific action or task. This interconnection is of two types:

1. Parallel connection
2. Series connection

Series Connected Electronic System

The system formed by the series connection of the individual blocks is called as "Series connected system". The output of the system formed by series connection of the two blocks is equal to the multiplication of the outputs of two individual blocks.

Here a complex system is formed by connecting the two individual blocks, G_1 and G_2.

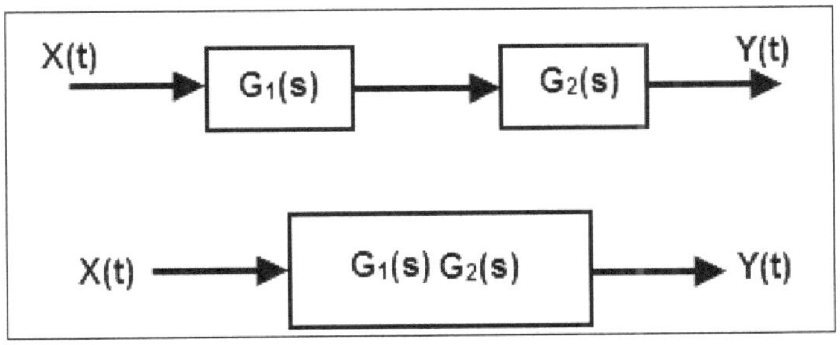

If the output of the system first system is G1 and the output of the system is G2 then the output of the series connected system is equal to Y (t) = $G_1(s)*G_2(s)$.

If the system has 3 blocks A, B, C connected in series, then the output of the series connected system is equal to Y = A*B*C. We can apply this for any number of blocks in series.

Example: a micro phone connected with amplifier and speaker.

Parallel Connected Electronic System

The system formed by the parallel connection of the individual blocks is called as "Parallel connected system". The output of the system formed by series connection of the two blocks is equal to the sum of the outputs of two individual blocks.

Here a complex system is formed by connecting the two individual blocks, G_1 and G_2.

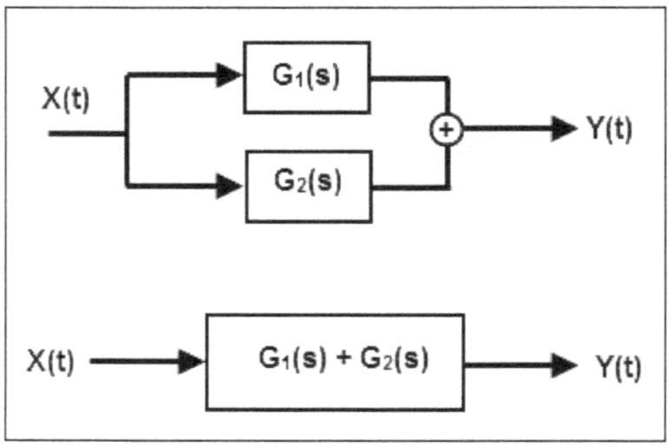

If the output of the system first system is G_1 and the output of the system is G_2 then the output of the parallel connected system is equal to Y (t) = G_1(s) + G_2(s).

If the system has 3 blocks A, B, C connected in series, then the output of the series connected system is equal to Y = A + B + C. We can apply this for any number of blocks in parallel.

Example: several micro-phones connected with single amplifier and to a single speaker system.

Electronic Feedback Systems

Feedback is another type of interconnection of systems. Feedback loop means connecting some part of output value connecting back to the input. The electronic systems having a connected feedback loop are called "Closed feedback system".

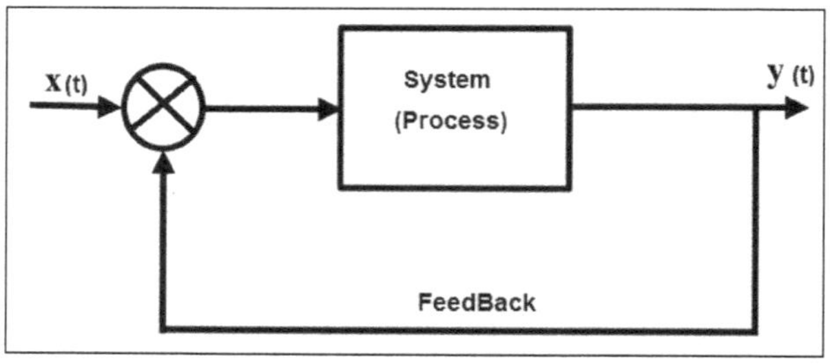

The feedback loop allows the system to check errors and avail it to auto correct the errors occurred in its output. So these systems are more stable than open loop systems (which doesn't contain any feedback loop).

Based on the nature of the feedback loop, the closed loop systems classified as positive feedback systems and negative feedback systems.

The system in which the feedback loop increases the level of output of the system is known as "Positive feedback electronic systems". Similarly, the system in which the feedback loop decreases the level of output of the system is known as "Negative feedback electronic systems".

Example: An automatic heating system is an example of the simple feedback system.

Linear Time Invariant Systems

Linear time-invariant systems (LTI systems) are a class of systems used in signals and systems that are both linear and time-invariant. Linear systems are systems whose outputs for a linear combination of inputs are the same as a linear combination of individual responses to those inputs. Time-invariant systems are systems where the output does not depend on when an input was applied. These properties make LTI systems easy to represent and understand graphically.

LTI systems are superior to simple state machines for representation because they have more memory. LTI systems, unlike state machines, have a memory of past states and have the ability to predict the future. LTI systems are used to predict long-term behavior in a system. So, they are often used to model systems like power plants. Another important application of LTI systems is electrical circuits. These circuits, made up of inductors, transistors, and resistors, are the basis upon which modern technology is built.

1. Dynamical System: Output depends on present and past values of input.

Examples:

$$y(t) = \int_{t-3}^{t} u^3(t_1)\,dt_1$$

$$y(t) = u(t) + \sum_{N=1}^{N} u(t - n\delta)$$

2. Time Invariance: Systems are time invariant if their behavior and characteristics do not vary over time. In other words, if the input to a system is time shifted the resulting output experiences an identical time shift.

The system u(t) → y(t) is invariant if u(t + τ) → y(t + τ) for all τ.

Check process for time invariance:

Step 1: Replace u (t) by u (t + τ)

Step 2: Replace y (t) by y (t + τ)

Step 3: Are the results from steps 1 and 2 equal?

Example: $y(t) = \left[u(t)^{3/4}\right]$ Clearly Time Invariant.

Example: $y(t) = \int_{o}^{t} \sqrt{u(t_1)}\,dt_1$ Check Time invariance:

Plug in t + τ for t on RHS:

$$\int_{0}^{t} \sqrt{u(t_1 + \tau)}\,dt_1 = \int_{\tau}^{t+\tau} \sqrt{u(\zeta)}\,d\zeta$$

Plug in $t+\tau$ for t on LHS: Note the indices. The integral is from zero to t, and we are simply replacing the occurrences of t with $t + \tau$ – not shifting the indices as in step 1 – thus the lower index does not change.

$$y(t+\tau)=\int_0^{t+\tau}\sqrt{u(t_1)}dt_1$$

They are not equal, this system is not time invariant.

Example: $y(t)=\int_{t-5}^{t}u^4(t_1)dt_1$

Plug in $t+\tau$ for t on RHS:

$$\int_{t-5}^{t}u^4(t+\tau)dt_1=\int_{t-5+\tau}^{t+\tau}u^4(\zeta)d\zeta$$

Plug in $t+\tau$ for t on LHS:

$$y(t+\tau)=\int_{t-5+\tau}^{t+\tau}u^4(t_1)dt_1$$

Example is time invariant.

3. Linear System: A system in linear if the superposition of several input signals results in an output which is the same superposition of the output signals of the system for each individual input.

$$\text{Additive}: x_1(t)+x_2(t)\rightarrow y_1(t)+y_2(t)$$
$$\text{Scaling}: ax_1(t)\rightarrow ay_1(t)$$
$$\text{Combined}: a_1x_1(t)+a_2x_2(t)\rightarrow a_1y_1(t)+a_2y_2(t)$$

A system must satisfy both the additive and the scaling criteria for it to be considered linear.

Example: $y(t)=C\frac{du}{dt}$ Linear.

Example: $y(t)=\int_0^t u(t_1)dt_1$ Linear, but not time invariant.

Example: $y(t)=au^3(\text{t})$ is not linear but is time invariant.

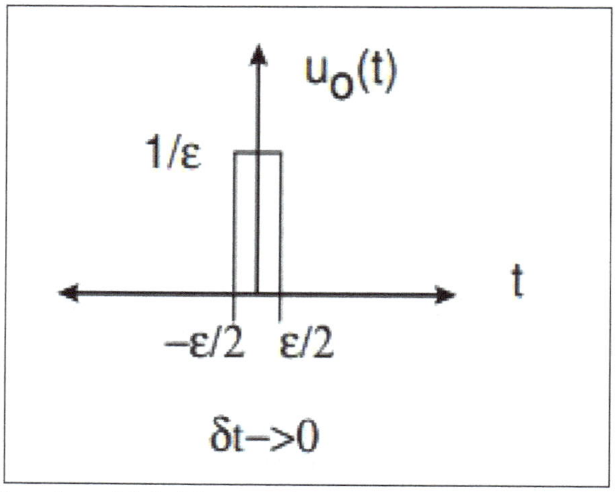

Figure: Delta function with height $1/\varepsilon$ between $-\varepsilon/2$ and $\varepsilon/2$ as ε goes to zero.

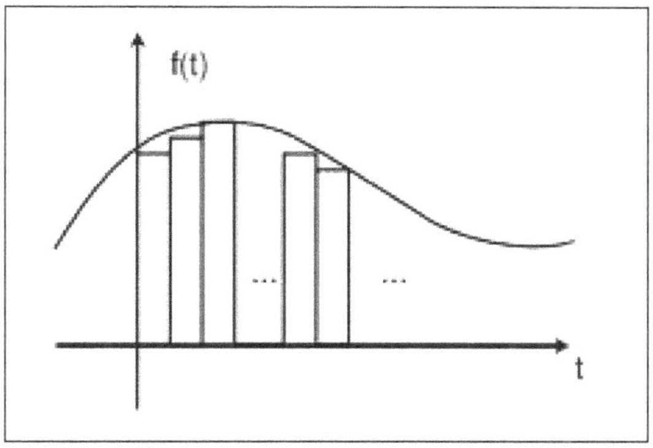

A function f(t) represented as a sum of scaled and time-shifted impulses.

4. Linear, Time-Invariant (LTI) Systems: Systems that satisfy both linear and time invariant criterion are considered LTI systems. The property of superposition makes LTI systems easier to analyze. By representing complex inputs as a superposition of basic signals, such as the impulse, we can then use superposition to determine the system output.

5. Impulse function: Representation of an time-continuous LTI system as a response to a unit impulse. A unit impulse is an idealization of a pulse which is so short that its duration is inconsequential for any real system.

$$u_o = \begin{matrix} 1/\in; & |t| \leq \varepsilon/2 \\ 0; & |t| > \varepsilon/2 \end{matrix}$$

The integral of an impulse from minus infinity to infinity is 1 and $u_o(t)$ is an even function: $u_o(t) = u_o(-t)$. Impulses can be scaled, shifted and summed to represent a function *f*(*t*).

6. Convolution: The convolution integral can be used to evaluate the response of a system to an input, $u(t)$.

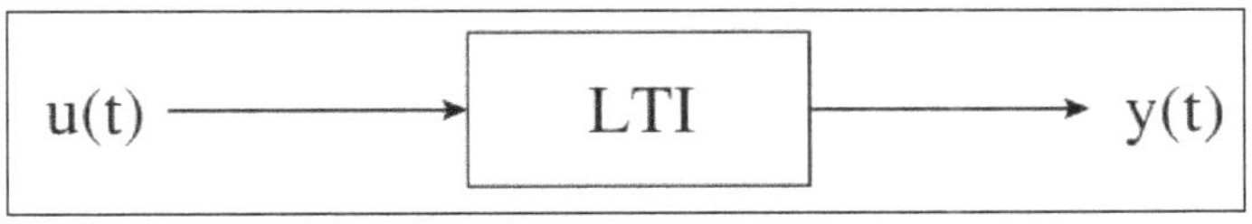

The impulse response of a linear time-invariant (LTI) system can be determined using the convolution integral.

$$y(t) = \int_{-\infty}^{\infty} u(t)h(t-\tau)dt$$

$$\int_{-\infty}^{\infty} u(t-\tau)h(\tau)d\tau$$

Where *h*(*t*) represents a unique function for each system and is called the Impulse Response.

$$\int_{-\infty}^{\infty} f(t)u_o(t)dt = lim\ \varepsilon \rightarrow 0 \int_{-\varepsilon/2}^{\varepsilon/2} f(t)u_o(t)dt$$

$$= f(o) \int_{-\varepsilon/2}^{\varepsilon/2} u_o(t)dt$$

$$= f(o)$$

The convolution integral holds for a time-shifted input:

$$\int_{-\infty}^{\infty} f(t)u_o(t-\zeta)dt = f(\zeta)$$

This gives us $f(t)$ about $\mathrm{t} = \zeta + / - \varepsilon / 2$.

Causality

A causal system responds only after being excited. All physical systems are causal. For a causal LTI system, the impulse response, $h(\mathrm{t})$ is zero for all $t<0$ since the input $u(t)=0$ before time zero.

$$\begin{aligned} y(\mathrm{t}) &= \int_{-\infty}^{\infty} u(\tau)h(t-\tau)d\tau \\ &= \int_{0}^{\infty} u(\tau)h(t-\tau)d\tau \\ &= \int_{0}^{t} u(\tau)h(t-\tau)d\tau \text{ since } h(\mathrm{t}-\tau)=0 \text{ for } \mathrm{t}-\tau<0 \end{aligned}$$

Impulse Response of a Linear System

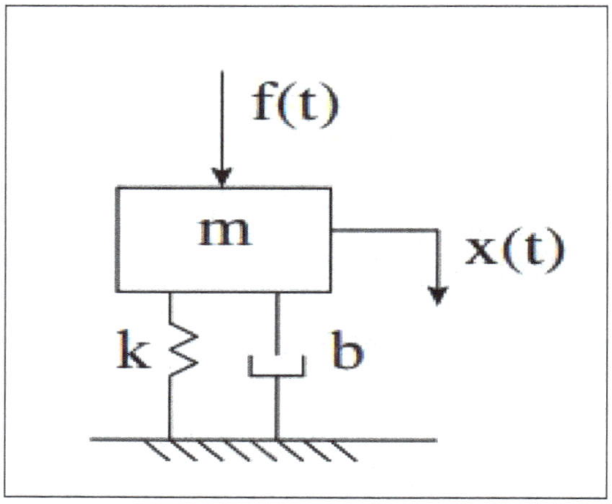

Mass-spring-dashpot system

Example of Mass-spring-dashpot

$$m\ddot{x} + b\dot{x} + kx = f(t)$$

(1) System starts at rest:

$$\begin{aligned} x(0) &= 0 \\ \dot{x}(0) &= 0 \end{aligned}$$

(2) Apply an impulse force $f(t) = u_o(t)$ at time zero and integrate over the duration of the impulse.

$$\int_{-\varepsilon/2}^{\varepsilon/2} \{m\ddot{x} + b\dot{x} + kx\} dt = \int_{-\varepsilon/2}^{\varepsilon/2} \{f(t)\}\, dt = 1$$

Due to the initial conditions we are left with:

$$m\left\{ \dot{x}\left(0^+\right) - \dot{x}\left(0^-\right)\right\} = 1$$

The term $\dot{x}(0^-)$ is zero since there is no motion before time zero and we are left with the velocity just after the force is applied:

$$\dot{x}\left(0^+\right) = \frac{1}{m}$$

(3) For times after the initial impulse force the problem becomes:

$$m\ddot{x} + b\dot{x} + kx = 0$$

$$x(0) = 0$$

$$\dot{x}\left(0^+\right) = \frac{1}{m}$$

From differential equations we have the solution to this system of equations:

$$x(t) \;=\; C_1 e^{s_1 t} + C_2 e^{s_2 t}$$

Plugging this into the equation $\dot{x}\left(0^+\right) = \frac{1}{m}$ we are left to solve for $s_{1,2}$.

$$ms^2 + bs + k = 0$$

$$s_{1,2} = -\frac{b}{2m} \pm \frac{\sqrt{b^2 - 4km}}{2m}$$

Let $\delta = \frac{b}{2m}$ and $\omega = \sqrt{\frac{k}{m} - \frac{b^2}{4m}}$

Assume $b^2 < 4km$

$$x(0) = C_1 + C_2 = 0 \rightarrow C_1 = -C_2$$

$$\dot{x}(0) = C_1 s_1 + C_2 s_2 = \frac{1}{m} = C_1\left(s_1 - s_2\right)$$

$$s_1 - s_2 = \delta + i\omega_d - \left(\delta - i\omega_d\right)$$

$$C = \frac{1}{2im\omega_d}$$

(4) Solution due to the impulse force input: Impulse Response

$$x(t)=\frac{1}{2im\omega_d}\left\{e^{-\delta t}\left(e^{i\omega_d t}-e^{-i\omega_d t}\right)\right\}$$

$$\sin\omega_d t=\frac{e^{i\omega_d t}-e^{-i\omega_d t}}{2i}$$

$$h(t)=\begin{matrix}\frac{1}{mw_d}e^{-\delta t}\sin w_d t; & t\geq 0\\ 0; & t<0\end{matrix}$$

Fourier Series

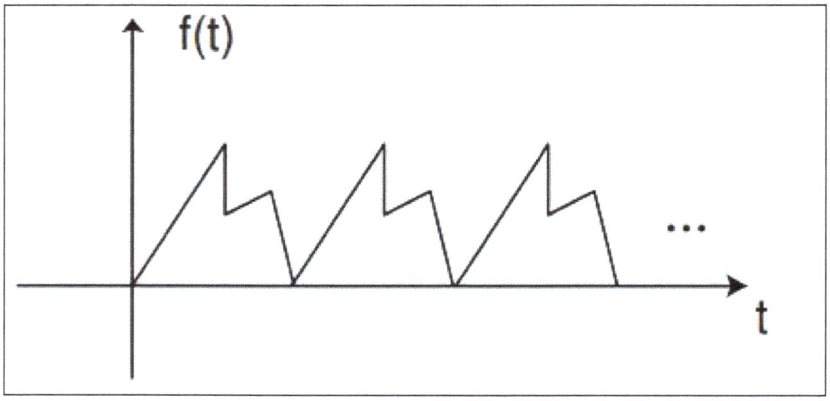

Periodic Signal

Fourier series can be used to represent a periodic signal as a summation of sines and cosine.

$$f(t)=A_o+\sum_{n=1}^{\infty}\left\{\mathrm{A}_n\cos(n\omega_o t)+b_n\sin(n\omega_o t)\right\}$$

Where $\omega_o=2\pi/T$ is the fundamental frequency and the coefficients are:

$$A_o=\frac{1}{T}\int_0^t f(t)\,dt$$

$$A_n=\frac{2}{T}\int_0^t f(t)\cos(n\omega_o t)\,dt$$

$$B_n=\frac{2}{T}\int_0^t f(t)\,\sin(n\omega_o t)\,dt$$

The fourier series can be written compactly in complex notation.

$$f(t)=\sum_{-\infty}^{\infty}\tilde{C}_n e^{in\omega_o t}$$

$$\tilde{C}_n = \frac{1}{T}\int_{-\infty}^{\infty} f(\mathrm{t})\,\mathrm{e}^{-in\omega_o t}\,dt$$

We can use fourier series as a representation of of a periodic absolutely integrable functions *f*(*t*). An absolutely integral function is one whos integral converges when between minus and plus infinity or has a finite number of discontinuities that can be integrated around.

$$\int_{-\infty}^{\infty} |x(t)|dt < \infty$$

Fourier Transform

Fourier transform (FT) converts a function of time into a function of frequency. The inverse fourier transform (IFT) reverts the function in frequency space back to time space.

- Fourier Transform

$$\tilde{f}(\omega) = \int_{-\infty}^{\infty} f(t)e^{-i\omega t}dt \equiv \text{ Fourier Transform}$$

- Inverse Fourier Transform

$$f(t) = \frac{1}{2\pi}\int_{-\infty}^{\infty} \tilde{f}(\omega)e_{i\omega t}d\omega \equiv \text{Inverse Fourier Transform}$$

Example: Fourier transform of $x(t)$:

$$x(t) = 1; \quad |t| \leq T_1$$

$$0; \quad |t| > T_1$$

$$\tilde{X}(t) = \int_{-T_1}^{T_1} e^{-i\omega t}(1)dt = \frac{2\sin\omega T_1}{\omega}$$

Example: Inverse fourier transform of $\tilde{X}(t)$:

$$\tilde{X}(w) = 1; \quad |\omega| \leq \mathrm{W}_1$$
$$0; \quad |\omega| > W_1$$

$$\tilde{X}(t) = \frac{1}{2\pi}\int_{-W_1}^{W_1} \tilde{X}(t)e^{iwt}d\omega = \frac{\sin W_1 t}{\pi t}$$

The similarity between the two functions in examples 1 and 2. Parseval's theorem explains that there exists a dual pair of functions with time and frequency interchanged. A symmetric pair of functions.

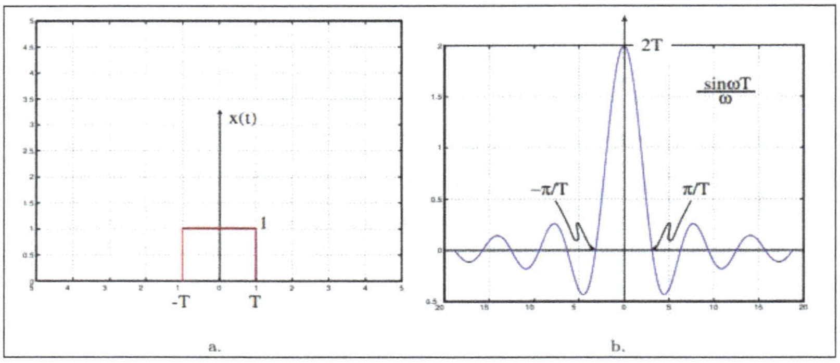

(a) Function of time: x(t); (b) Fourier transform of x(t) in frequency space.

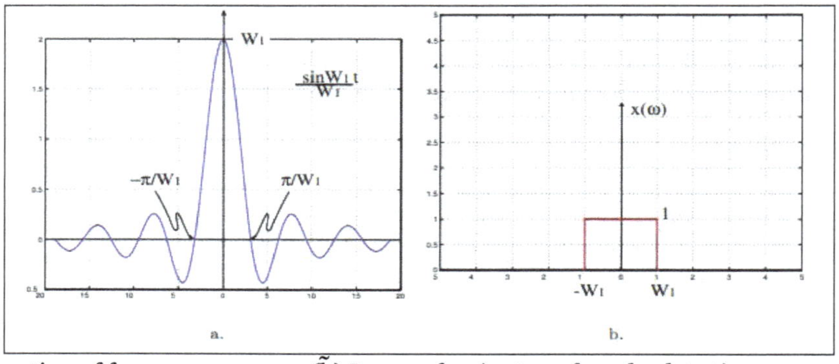

(a) Function of frequency (b) Inverse fourier transform back to time space

Convolution and the Fourier Transform

For LTI systems, the fourier transform turns the convolution integral into simple multiplication.

$$\tilde{Y}(\omega) = FT\left\{\int_{-\infty}^{\infty} u(\tau)h(t-\tau)d\tau\right\}$$

$$= \int_{-\infty}^{\infty}\int_{-\infty}^{\infty}\left(u(\tau)h(t-\tau)d\tau\right)e^{-i\omega t}dt$$

$$e^{-i\omega t} = e^{-i\omega(t+\tau-\tau)}$$

Let $t-\tau = t_1$ and $dt = dt_1$, then we get:

$$\tilde{Y}(\omega) = \int_{-\infty}^{\infty}\int_{-\infty}^{\infty} u(\tau)h(t_1)\,e^{-i\omega(t_1+\tau)}d\tau\delta t_1$$

$$= \underbrace{\left\{\int_{-\infty}^{\infty} h(t_1)e^{-i\omega t_1}dt_1\right\}}_{\tilde{h}(\mathrm{w})} \underbrace{\left\{\int_{-\infty}^{\infty} u(\tau)e^{-i\omega\tau}d\tau\right\}}_{\tilde{u}(\mathrm{w})}$$

$$\tilde{Y}(\omega) = \tilde{h}(\mathrm{w})\cdot\tilde{u}(\mathrm{w}).$$

Where $\tilde{h}(\mathrm{w})$ is the fourier transform of the impulse response and is referred to as the Transfer Function, $H(\mathrm{w})$.

$$\tilde{Y}(\omega) = \tilde{H}(\mathrm{w}) \cdot \tilde{u}(\mathrm{w})$$

Recap of Fourier Transforms

- Convolution: $y(t) = h(t) * u(t)$
- Multiplication: $\tilde{Y}(\omega) = H(\omega) \cdot \tilde{u}(\omega)$
- Linearity: If $x(t) \rightarrow \tilde{x}(\omega)$ and $y(t) \rightarrow \tilde{y}(\omega)$ then $ax(t) + by(t) \rightarrow a\tilde{x}(\omega) + b\tilde{y}(\omega)$.

LTI and Fourier Transforms

To evaluate a LTI system you can use the fourier transform and convolution to find the output y(t) given the input and the transfer function.

- $u(t) \rightarrow \tilde{u}(\omega)$ take the FT of the input.
- $\tilde{y}(\omega) = H(\omega)\tilde{u}(\omega)$ convolve the FT of the input and the transform function.
- $\tilde{y}(\omega) \rightarrow y(t)$ take the inverse FT to find the system output.

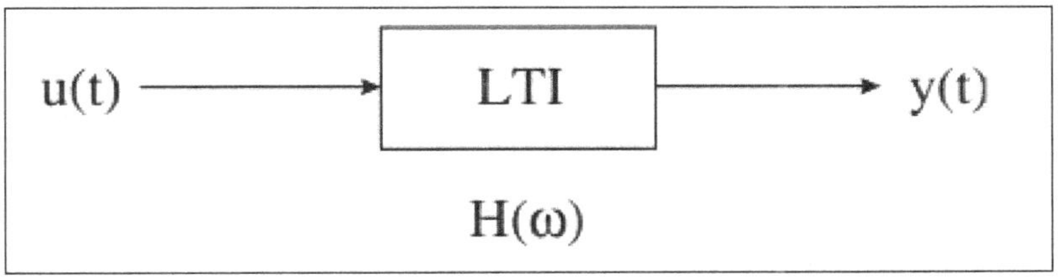

LTI System Block Diagram

For a given a harmonic input $u(t)$ and transfer function $H(\omega)$ we can easily write the output of our LTI system in terms of the amplitude and frequency of the input and the amplitude and argument (phase) of the transfer function.

$$u(t) = u_o \cos(\omega_o t + \psi)$$

$$y(t) = y_o \cos(\omega_o t + \psi + \psi_1)$$

Where $y_o = u_o |H(\omega_o)|$ is the amplitude of the response and $\psi_1 = arg\{H(w_o)\}$ is the phase shift of the response from the input.

From complex math we can write a function H(ω) in terms of its amplitude, |H(ω)|, and its argument, $\angle H(\omega)$ as follows:

$$H(\omega) = |H(\omega)| e^{i\angle H(\omega)}$$

Now lets look at the real part of a complex function:

$$u(t) = Re\{u_o e^{i(\omega_o t + \psi)}\} = Re\{\tilde{u}_o e^{i\omega_o t}\}$$

Where $\tilde{u}_o = u_o e^{i\psi}$. Taking the convolution $y(t) = h(t) * u(t)$ we have:

$$\begin{aligned} y(t) &= \int_{-\infty}^{\infty} h(\tau) Re\left\{\tilde{u}_o e^{i\omega_o(t-\tau)}\right\} d\tau \\ &= Re\left\{\left\{\int_{-\infty}^{\infty} h(\tau) e^{-i\omega_o\tau} d\tau\right\} \tilde{u}_o e^{i\omega_o t}\right\} \\ &= Re\left\{H(\omega_o) \tilde{u}_o e^{i\omega_o t}\right\} \end{aligned}$$

Where $H(\omega_o) = |H(\omega_o)| e^{i\angle H(\omega_o)}$.

Since we are only interested in the real part of y(t) (the input was cosine) we have:

$$y(t) = u_o |H(\omega_o)| \cos(\omega_o t + \psi + \angle\{H(\omega_o)\}$$

This process can be extended in the same fashion for an input of sine. However instead of the real component we look at the imaginary component of $e^{i\omega t} = \cos(\omega t) + i\sin(\omega t)$ which is $\sin(\omega t)$.

Properties of LTI Systems

LTI systems are those that are both linear and time-invariant.

Linear systems have the property that the output is linearly related to the input. Changing the input in a linear way will change the output in the same linear way. So if the input $x_1(t)$ produces the output $y_1(t)$ and the input $x_2(t)$ produces the output $y_2(t)$, then linear combinations of those inputs will produce linear combinations of those outputs. The input $(x_1(t) + x_2(t))$ will produce the output $(y_1(t) + y_2(t))$. Further, the input $(a_1 \cdot x_1(t) + a_2 \cdot x_2(t))$ will produce the output $(a_1 \cdot y_1(t) + a_2 \cdot y_2(t)$ for some constants a_1 and a_2.

In other words, for a system *T* over time *t*, composed of signals $x_1(t)$ and $x_2(t)$ with outputs $y_1(t)$ and $y_2(t)$,

$$T[a_1 x_1(t) + a_2 x_2(t)] = a_1 T[x_1(t)] + a_2 T[x_2(t)] = a_1 y_1(t) + a_2 y_2(t),$$

Where a_1 and a_2 are constants.

Further, the output of a linear system for an input of 0 is also 0.

Time-invariant systems are systems where the output for a particular input does not change depending on when that input was applied. A time-invariant systems that takes in signal $x(t)$ and produces output $y(t)$ will also, when excited by signal $x(t+\sigma)$, produce the time-shifted output $y(t+\sigma)$.

Thus, the entirety of an LTI system can be described by a single function called its impulse response. This function exists in the time domain of the system. For an arbitrary input, the output of an LTI system is the convolution of the input signal with the system's impulse response.

Conversely, the LTI system can also be described by its transfer function. The transfer function is the Laplace transform of the impulse response. This transformation changes the function from the time domain to the frequency domain. This transformation is important because it turns differential equations into algebraic equations, and turns convolution into multiplication. In the frequency domain, the output is the product of the transfer function with the transformed input. The shift from time to frequency is illustrated in the following image:

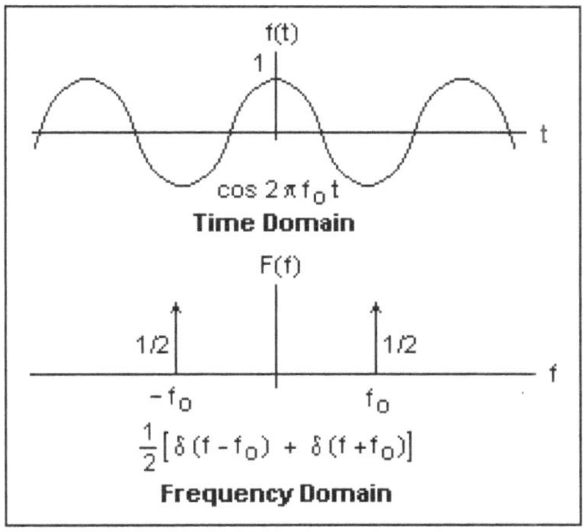

In addition to linear and time-invariant, LTI systems are also memory systems, invertible, casual, real, and stable. That means they have memory, they can be inverted, they depend only on current and past events, they have fully real inputs and outputs, and they produce bounded output for bounded input.

Because of the properties of LTI systems, the general form of an LTI system with output $y[n]$ and input $x[n]$ at time n, and constants c_k and d_j is defined a.

$$y[n] = c_0 y[n-1] + c_1 y[n-2] + \ldots + c_{k-1} y[n-k] + d_0 x[n] + d_1 x[n-1] + \ldots + d_j x[n-j].$$

The state of this system depends on the previous k output values and jj input values. Because of the linearity property, the output at time n is just a linear combination of the previous outputs, previous inputs, and current input.

Further, if a string of LTI systems are cascaded together, the output of that new system does not depend on the order in which the systems were cascaded. This property follows from the associative property and the commutative property.

We can take the general form of the LTI system, and write it as an operator equation, and with some manipulation we can turn it into a useful formula:

$$\begin{aligned} Y &= c_0 RY + c_1 R^2 Y + \cdots + c_{k-1} R^k Y + d_0 X + d_1 RX + \cdots + d_j R^j X \\ &= Y(c_0 R + c_1 R^2 + \cdots + c_{k-1} R^k) + X(d_0 + d_1 R + \cdots + d_j R^j). \end{aligned}$$

This is the same equation as:

$$Y(1 - c_0 R - c_1 R^2 - \ldots - c_{k-1} R^k) = X(d_0 + d_1 R + \cdots + d_j R^j).$$

We can then do some division to create an equation that describes the quotient of the output signal and the input signal:

$$\frac{Y}{X}=\frac{d_0+d_1R+\cdots+d_jR_j}{1-c_0R-c_1R^2-\ldots-c_{k-1}R^k}.$$

This is the system function of the LTI system, and it is typically written as the polynomial:

$$\frac{Y}{X}=\frac{b_0+b_1R+b_2R^2+\cdots}{a_0+a_1R+a_2R^2+\cdots}.$$

The numerator and the denominator are polynomials in R, the delay variable.

Control System

A control system is a system of devices or set of devices, that manages, commands, directs or regulates the behavior of other devices or systems to achieve desired results. In other words, the definition of a control system can be simplified as a system, which controls other systems. As the human civilization is being modernized day by day the demand for automation is increasing accordingly. Automation highly requires control of devices.

In recent years, control systems have played a central role in the development and advancement of modern technology and civilization. Practically every aspect of our day-to-day life is affected less or more by some control system. A bathroom toilet tank, a refrigerator, an air conditioner, a geezer, an automatic iron, an automobile all are control system. These systems are also used in industrial processes for more output. We find control systems in the quality control of products, weapons system, transportation systems, power system, space technology, robotics and many more. The principles of control theory are applicable to both engineering and non-engineering field.

Features of a Control System

The main feature of a control system is that there should be a clear mathematical relationship between input and output of the system. When the relation between input and output of the system can be represented by a linear proportionality, the system is called a linear control system. Again when the relationship between input and output cannot be represented by single linear proportionality, rather the input and output are related by some non-linear relation, the system is referred to as a non-linear control system.

Requirements of a Good Control System

- Accuracy: Accuracy is the measurement tolerance of the instrument and defines the limits of the errors made when the instrument is used in normal operating conditions. Accuracy can be improved by using feedback elements. To increase the accuracy of any control system error detector should be present in the control system.

- Sensitivity: The parameters of a control system are always changing with the change in surrounding conditions, internal disturbance or any other parameters. This change can be expressed in terms of sensitivity. Any control system should be insensitive to such parameters but sensitive to input signals only.
- Noise: An undesired input signal is known as noise. A good control system should be able to reduce the noise effect for better performance.
- Stability: It is an important characteristic of the control system. For the bounded input signal, the output must be bounded and if the input is zero then output must be zero then such a control system is said to be a stable system.
- Bandwidth: An operating frequency range decides the bandwidth of the control system. Bandwidth should be as large as possible for the frequency response of good control system.
- Speed: It is the time taken by the control system to achieve its stable output. A good control system possesses high speed. The transient period for such system is very small.
- Oscillation: A small numbers of oscillation or constant oscillation of output tend to indicate the system to be stable.

PID Controller

PID Controller is a most common control algorithm used in industrial automation & applications and more than 95% of the industrial controllers are of PID type. PID controllers are used for more precise and accurate control of various parameters.

Most often these are used for the regulation of temperature, pressure, speed, flow and other process variables. Due to robust performance and functional simplicity, these have been accepted by enormous industrial applications where a more precise control is the foremost requirement.

A combination of proportional, integral and derivative actions is more commonly referred as PID action and hence the name, PID (Proportional-Integral-Derivative) controller. These three basic coefficients are varied in each PID controller for specific application in order to get optimal response.

It gets the input parameter from the sensor which is referred as actual process variable. It also accepts the desired actuator output, which is referred as set variable, and then it calculates and combines the proportional, integral and derivative responses to compute the output for the actuator.

Consider the typical control system shown in above figure in which the process variable of a process has to be maintained at a particular level. Assume that the process variable is temperature (in centigrade). In order to measure the process variable (i.e., temperature), a sensor is used (let us say an RTD).

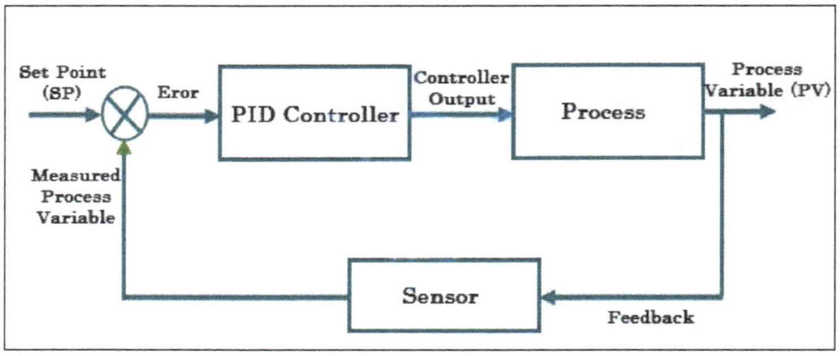

A set point is the desired response of the process. Suppose the process has to be maintained at 80 degree centigrade, and then the set point is 80 degree centigrade. Assume that the measured temperature from the sensor is 50 degree centigrade, (which is nothing but a process variable) but the temperature set point is 80 degree centigrade.

This deviation of actual value from the desired value in the PID control algorithm causes to produce the output to the actuator (here it is a heater) depending on the combination of proportional, integral and derivative responses. So the PID controller continuously varies the output to the actuator till the process variable settle down to the set value. This is also called as closed loop feedback control system.

Working of PID Controller

With the use of low cost simple ON-OFF controller only two control states are possible, like fully ON or fully OFF. It is used for limited control application where these two control states are enough for control objective. However oscillating nature of this control limits its usage and hence it is being replaced by PID controllers.

PID controller maintains the output such that there is zero error between process variable and set point/desired output by closed loop operations. PID uses three basic control behaviors that are explained below.

P- controller

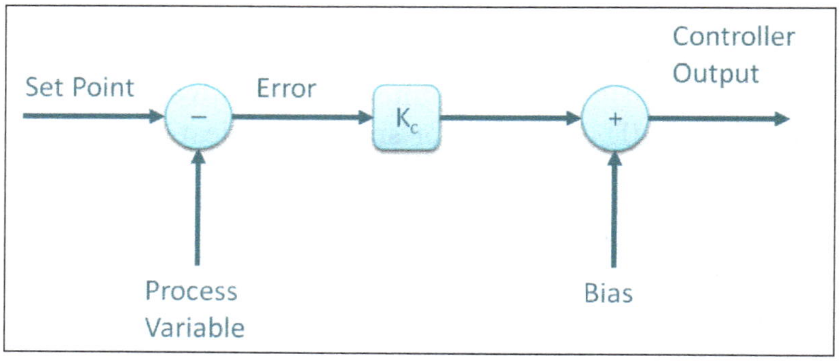

P-controller

Proportional or P-controller gives output which is proportional to current error e (t). It compares desired or set point with actual value or feedback process value. The resulting error is multiplied with proportional constant to get the output. If the error value is zero, then this controller output is zero.

This controller requires biasing or manual reset when used alone. This is because it never reaches the steady state condition. It provides stable operation but always maintains the steady state error. Speed of the response is increased when the proportional constant Kc increases.

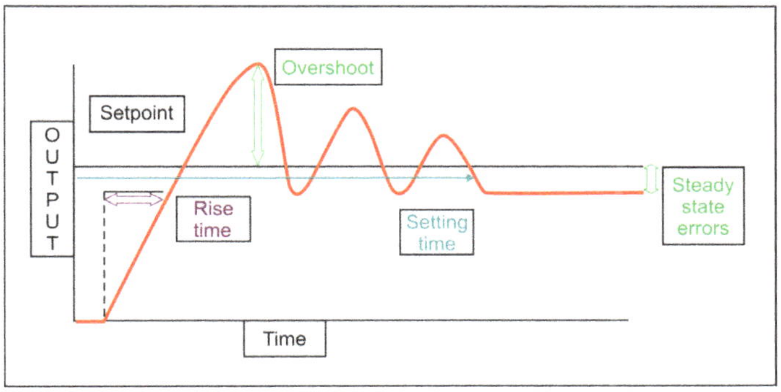

I-controller

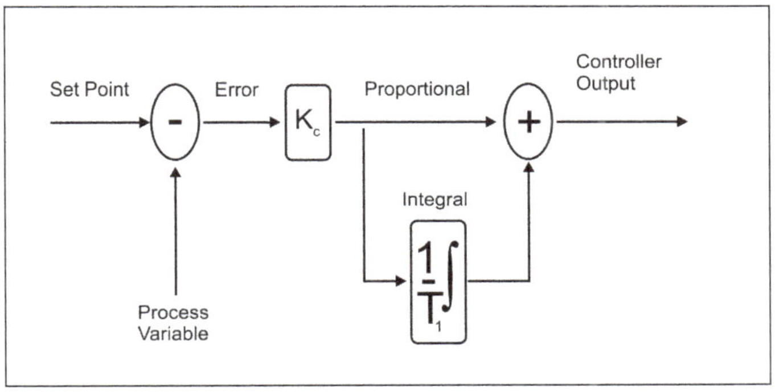

PI controller

Due to limitation of p-controller where there always exists an offset between the process variable and set point, I-controller is needed, which provides necessary action to eliminate the steady state error. It integrates the error over a period of time until error value reaches to zero. It holds the value to final control device at which error becomes zero.

Integral control decreases its output when negative error takes place. It limits the speed of response and affects stability of the system. Speed of the response is increased by decreasing integral gain Ki.

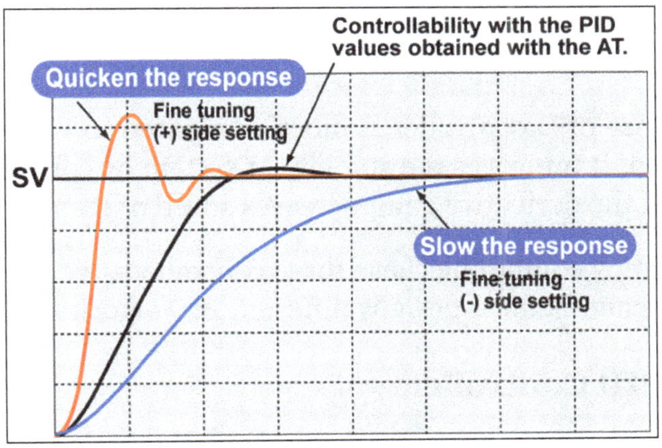

PI Controller Response

In figure, as the gain of the I-controller decreases, steady state error also goes on decreasing. For most of the cases, PI controller is used particularly where high speed response is not required.

While using the PI controller, I-controller output is limited to somewhat range to overcome the integral wind up conditions where integral output goes on increasing even at zero error state, due to nonlinearities in the plant.

D-controller

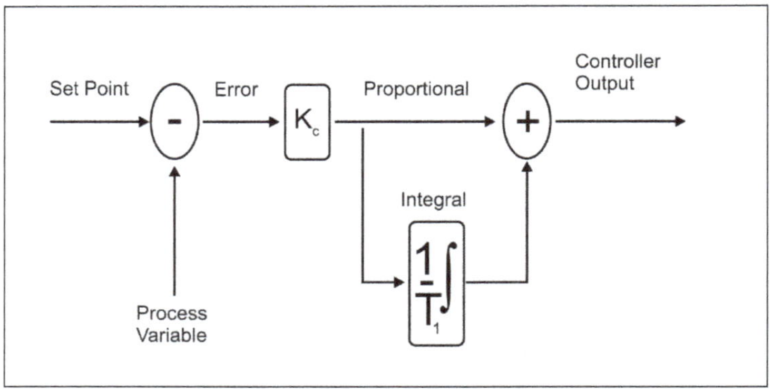

PID controller

I-controller doesn't have the capability to predict the future behavior of error. So it reacts normally once the set point is changed. D-controller overcomes this problem by anticipating future behavior of the error. Its output depends on rate of change of error with respect to time, multiplied by derivative constant. It gives the kick start for the output thereby increasing system response.

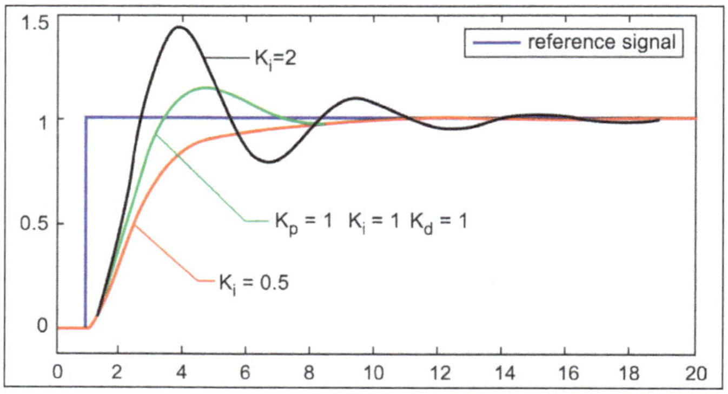

PID Controller Response

In the above figure response of D controller is more, compared to PI controller and also settling time of output is decreased. It improves the stability of system by compensating phase lag caused by I-controller. Increasing the derivative gain increases speed of response.

So finally we observed that by combining these three controllers, we can get the desired response for the system. Different manufactures designs different PID algorithms.

Tuning Methods of PID Controller

Before the working of PID controller takes place, it must be tuned to suit with dynamics of the process to be controlled. Designers give the default values for P, I and D terms and these values couldn't give the desired performance and sometimes leads to instability and slow control performances. Different types of tuning methods are developed to tune the PID controllers and require much attention from the operator to select best values of proportional, integral and derivative gains. Some of these are given below.

1. Trial and Error Method: It is a simple method of PID controller tuning. While system or controller is working, we can tune the controller. In this method, first we have to set Ki and Kd values to zero and increase proportional term (Kp) until system reaches to oscillating behavior. Once it is oscillating, adjust Ki (Integral term) so that oscillations stops and finally adjust D to get fast response.

2. Process reaction curve technique: It is an open loop tuning technique. It produces response when a step input is applied to the system. Initially, we have to apply some control output to the system manually and have to record response curve.

After that we need to calculate slope, dead time, rise time of the curve and finally substitute these values in P, I and D equations to get the gain values of PID terms.

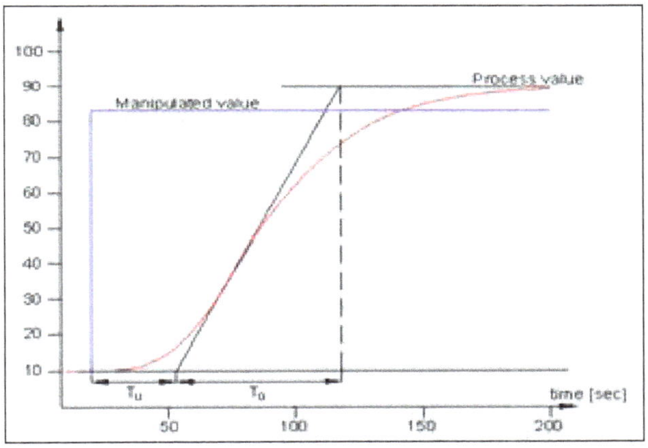

Process reaction curve

3. Zeigler-Nichols method: Zeigler-Nichols proposed closed loop methods for tuning the PID controller. Those are continuous cycling method and damped oscillation method. Procedures for both methods are same but oscillation behavior is different. In this, first we have to set the p-controller constant, Kp to a particular value while Ki and Kd values are zero. Proportional gain is increased till system oscillates at constant amplitude.

Gain at which system produces constant oscillations is called ultimate gain (Ku) and period of oscillations is called ultimate period (Pc). Once it is reached, we can enter the values of P, I and D in PID controller by Zeigler-Nichols table depends on the controller used like P, PI or PID, as shown in table.

	K_c	T_I	T_D
P	$K_u/2$		
PI	$K_u/2.2$	$P_u/1.2$	
PID	$K_u/1.7$	$P_u/2$	$P_u/8$

PID Controller Structure

PID controller consists of three terms, namely proportional, integral and derivative control. The combined operation of these three controllers gives control strategy for process control. PID controller manipulates the process variables like pressure, speed, temperature, flow, etc. Some of the applications use PID controllers in cascade networks where two or more PID's are used to achieve control.

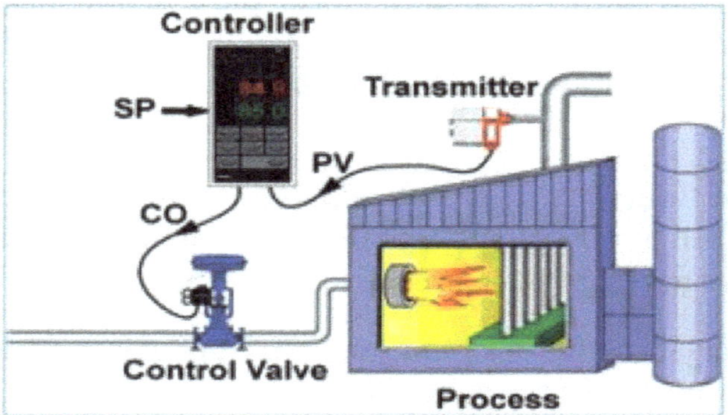

PID Controller Structure

Above figure shows structure of PID controller. It consists of a PID block which gives its output to process block. Process/plant consists of final control devices like actuators, control valves and other control devices to control various processes of industry/plant.

Feedback signal from the process plant is compared with a set point or reference signal u(t) and corresponding error signal e(t) is fed to the PID algorithm. According to the proportional, integral and derivative control calculations in algorithm, the controller produces combined response or controlled output which is applied to plant control devices.

All control applications don't need all the three control elements. Combinations like PI and PD controls are very often used in practical applications.

References

- Electronic-system: electronicshub.org, Retrieved 7 March, 2019
- Linear-time-invariant-systems: brilliant.org, Retrieved 15 June, 2019
- Continuous-lti-system-example, linear-time-invariant-systems: brilliant.org, Retrieved 18 January, 2019
- Control-system-closed-loop-open-loop-control-system: electrical4u.com, Retrieved 28 May, 2019
- What-is-pid-controller-how-it-works: electricaltechnology.org, Retrieved 8 February, 2019
- The-working-of-a-pid-controller: elprocus.com, Retrieved 9 April, 2019

5

System Theory and Function

System function or transfer function is used to show the relationship between the input signal and the output signal for all the possible input values. Some of the methods for obtaining a transfer function are block diagram method and signal flow graphs. This chapter discusses in detail these theories and methodologies related to system theory and functions.

System Theory

Systems theory is a set of theoretical concepts used to understand a wide variety of phenomena in terms of a model called a system.

The obvious inference that system theory deals with systems does not shed much light on it, since all branches of science are concerned with systems of one kind or another. The distinguishing characteristic of system theory is its generality and abstractness, its concern with the mathematical properties of systems and not their physical form.

System theory is not as yet a well-crystallized body of concepts and techniques which set it sharply apart from other better established fields of science. Indeed, there is considerable overlap and interrelation between system theory on the one hand and circuit theory, information theory, control theory, signal theory, operations research and systems engineering on the other. Yet, system theory has a distinct identity of its own which perhaps can be more clearly defined by listing its principal problems and areas.

Principal Problems of System Theory

1. System characterization (representation of input output relationships in mathematical form; transition from one mode of representation to another).
2. System classification (determination, on the basis of observation of input and output, of one among a specified class of systems to which the system under test belongs).
3. System identification (determination, on the basis of observation of input and output, of a system within a specified class of systems to which the system under test is equivalent; determination of the initial or terminal state of the system under test).
4. Signal representation (mathematical representation of a signal as a combination of elementary signals; mathematical description of signals).
5. Signal classification (determination of one among a specified set of classes of signals or patterns to which an observed signal belongs).

6. System analysis (determination of input-output relationships of a system from the knowledge of input-output relationships of each of its components).
7. System synthesis (specification of a system which has prescribed input-output relationships).
8. System control and programming (determination of an input to a given system which results in a specified or optimal performance).
9. System optimization (determination of a system within a prescribed class of systems which is best in terms of a specified performance criterion).
10. Learning and adaptation (problem of designing systems which can adapt to changes in environment and learn from experience).
11. Reliability (problem of synthesizing reliable systems out of less reliable components).
12. Stability and controllability (determination of whether a given system is stable or unstable, controllable-subject to specified constraints-or not controllable).

Principal Types of Systems

1. Linear, nonlinear.
2. Time-varying, time-invariant.
3. Discrete-time (sampled-data), continuous-time.
4. Finite-state, discrete-state, continuous-state.
5. Deterministic (non-random), probabilistic (stochastic).
6. Differential (characterized by integral-differential equations), no differential.
7. Small-scale, large-scale (large number of components).

State and State-space Techniques

Turing machine is a discrete time $=(t=0,\ 1,\ 2,\ \cdots)$ system with a finite number of states or internal configurations, which is subjected to an input having the form of a sequence of symbols (drawn from a finite alphabet) printed on a tape which can move in both directions along its length. The output of the machine at time t is an instruction to print a particular symbol in the square scanned by the machine at time t and to move in one or the other direction by one square. A key feature of the machine is that the output at time t and the state at time t+1 are determined by the state and the input at time t. Thus, if the state, the input and the output at time t are denoted by s_t, u, and y_t, respectively, then the operation of the machine is characterized by:

$$S_{t+l} = f(s_t, u_t), \quad t = 0, 1, 2, \ldots,$$

$$y_t = g(s_t,\ u_t)$$

Where f and g are functions defined on pairs of values of s_t and u_t. Note that above both imply that the output symbols from any initial time to on are determined by- the state at time to and the input symbols from time to on.

An important point about this representation, which was not pointed out by Turing, is that it is applicable not only to the Turing machine, but more generally to any discrete-time system. Furthermore, we shall presently see that it is a simple matter to extend $S_{t+l} = f(s_t, u_t)$, $t = 0, 1, 2, \ldots$, and $y_t = g(s_t, u_t)$ to systems having a continuum of states (i.e., continuous-state systems).

The characterization of a system by means of equations of the form $S_{t+l} = f(s_t, u_t)$, $t = 0, 1, 2, \ldots$, and $y_t = g(s_t, u_t)$ (to which we will refer as the Turing representation or, alternatively, as the state equations of a system) was subsequently employed by Shannon in his epoch-making paper on the mathematical theory of communication. Specifically, Shannon used $S_{t+l} = f(s_t, u_t)$, $t = 0, 1, 2, \ldots$, and $y_t = g(s_t, u_t)$ to characterize finite-state noisy channels, which are probabilistic systems in the sense that s_t and u_t determine not S_{t+1} and y_t, but their joint probability distribution. This implies that the system is characterized by both equations, with f and g being random functions, or, equivalently, by the conditional distribution function $p(s_{t+1}, y_t | s_t, u_t)$, where for simplicity of notation the same letter is used to denote both a random variable and a particular value of the random variable (e.g., instead of writing St for the random variable and s_t for its value, we use the same symbol, s_t, for both).

It was primarily Shannon's use of the Turing representation that triggered its wide application in the analysis and synthesis of discrete-state systems. Worthy of note in this connection is the important work of von Neumann on probabilistic logics, which demonstrated that it is possible, at least in principle, to build systems of arbitrarily high degree of reliability from unrealizable (probabilistic) components. Also worthy of note is the not-too-well-known work of Singleton [41 on the theory of nonlinear transducers, in which techniques for optimizing the performance of a system with quantized state space are developed. It should be remarked that the problem of approximating to a system having a continuum of states with a system having a finite or countable number of states is a basic and as vet unsolved problem in system theory. Among the few papers which touch upon this problem, those by Kaplan contain significant results for the special case of a differential5 system subjected to zero input. A qualitative discussion of the related problem of ε-approximation may be found in a paper by Stebakov.

There are two important notions that are missing or play minor roles in the papers cited above. These are the notions of equivalent states and equivalent machines which were introduced by Moore and, independently and in a somewhat restricted form, by Huffman. The theory developed by Moore constitutes a contribution of basic importance to the theory of discrete state systems and, more particularly, the identification problem.

So far, our discussion of the notion of state has been conducted in the context of discrete-state systems. In the case of a differential system, the state equations $S_{t+l} = f(s_t, u_t)$, $t = 0, 1, 2, \ldots$, and $y_t = g(s_t, u_t)$ assume the form:

$$\dot{s}(t) = f(s(t), u(t))$$
$$y(t) = g(s(t), u(t))$$

Where $\dot{s}(t) = d/dt\, s(t)$, and $s(t)$, $y(t)$ and $u(t)$ are vectors representing the state, the input and output of the system at time t. Under various guises, such equations [particularly for the case where $u(t) \equiv 0$] have long been used in the theory of ordinary differential equations, analytical dynamics, celestial mechanics, quantum mechanics, econometrics, and other fields.

State and System Equivalence

Suppose that we have two systems B_1 and B_2, with qi being a state of B, and q_2 being a state of B_2. As the term implies, q_i and q_2 are equivalent states if, for all possible input time-functions u, the response of B_1 to u starting in state q, is the same as the response of B_2 to u starting in state q_2. Following Moore, B_1 and B_2 are said to be equivalent systems if, and only if, to every state in B, there is an equivalent state in B_2, and vice versa. What is the significance of this definition? Roughly speaking, it means that if B_1 and B_2 are equivalent, then it is impossible to distinguish B, from B_2 by observing the responses of B , and B_2 to all possible inputs u, if the initial states of B, and B_2 are not known to the experimenter.

To illustrate, let us consider the two simple networks shown in figure. The state of B_1, is a vector with two components (which may be identified with the currents flowing through the two inductances); the state of B_2 is a scalar (the current through 2L). Nevertheless, by writing the state equations for B_1 and B_2, it is easy to verify that B, and B_2 are equivalent in the sense defined above, as well as in the more conventional sense of circuit theory.

On the other hand, consider the constant-resistance network shown in figure. Its input impedance is equal to unity at all frequencies. Does this mean that B_1, is equivalent to a unit resistor?

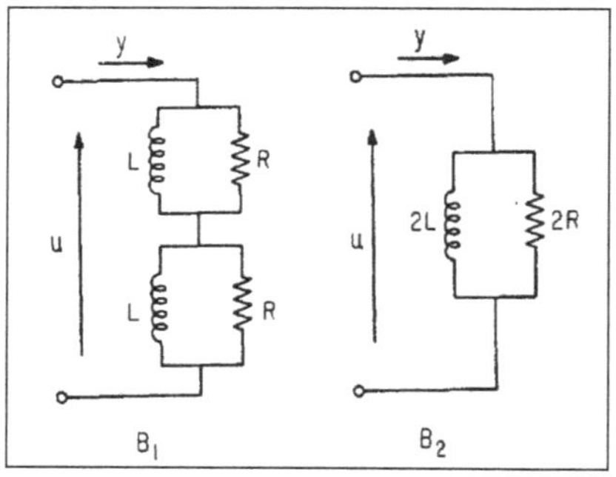

Equivalent Networks

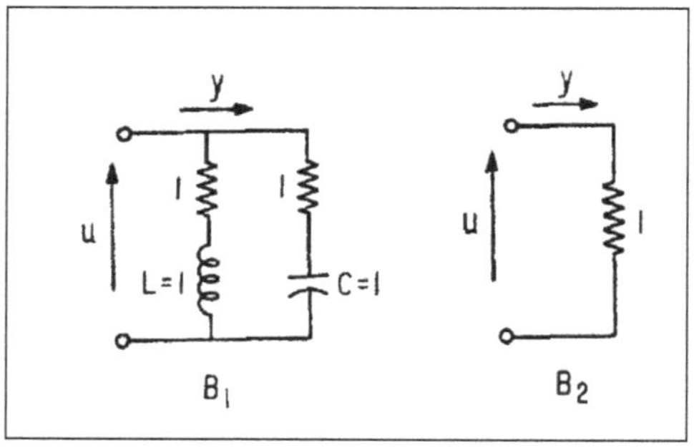

Ground state equivalent networks

The circuit theorist's answer to this question would be "yes,") since in circuit theory two networks are defined to be equivalent if their respective terminal impedances (or admittances) have the same values at all frequencies. By contrast, the system theorist would say "no," since there are states in B_1, say the state (0, 1) (where the first component is the voltage across C and the second component is the current through L), to which no equivalent state in the unit resistor can be found.8 However, we note that the unexcited state (ground state) of B_1, which is (0,0), is equivalent to the ground state of the resistor. Thus, we can assert that, although B_1 is not equivalent to the unit resistor, it is ground-state equivalent to the unit resistor. In plain words, this means that if B is initially unexcited, then it will behave like a unit resistor. On the other hand, if B is initially excited, then it may not behave like a unit resistor.

This simple example shows that the notion of equivalence in circuit theory has a narrower meaning than it does in system theory. M\lore specifically, the equivalence of two networks in the sense of system theory implies, but is not implied by, their equivalence in the sense of circuit theory. In effect, the notion of equivalence in circuit theory corresponds to the notion of ground-state equivalence in system theory.

System Function

System function is also Known as Transfer function.

A transfer function represents the relationship between the output signal of a control system and the input signal, for all possible input values. A block diagram is a visualization of the control system which uses blocks to represent the transfer function, and arrows which represent the various input and output signals.

For any control system, there exists a reference input known as excitation or cause which operates through a transfer operation (i.e. the transfer function) to produce an effect resulting in controlled output or response.

Thus the cause and effect relationship between the output and input is related to each other through a transfer function.

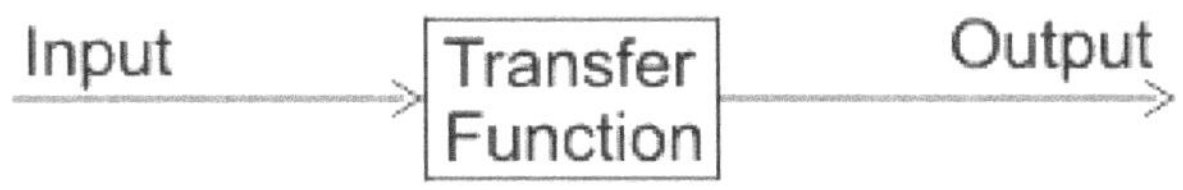

In a Laplace Transform, if the input is represented by R(s) and the output is represented by C(s), then the transfer function will be:

$$G(s) = \frac{C(s)}{R(s)} \Rightarrow R(s).G(s) = C(s)$$

That is, the transfer function of the system multiplied by the input function gives the output function of the system.

The transfer function of a control system is defined as the ratio of the Laplace transform of the output variable to Laplace transform of the input variable assuming all initial conditions to be zero.

$$G(s) = \frac{C(s)}{R(s)}$$

Procedure for determining the transfer function of a control system are as follows:

1. We form the equations for the system.
2. Now we take Laplace transform of the system equations, assuming initial conditions as zero.

3. Specify system output and input.
4. Lastly we take the ratio of the Laplace transform of the output and the Laplace transform of the input which is the required transfer function.

It is not necessary that output and input of a control system are of same category. For example, in electric motors the input is electrical signal whereas the output is mechanical signal since electrical energy required to rotate the motors. Similarly in an electric generator, the input is mechanical signal and the output is electrical signal, since mechanical energy is required to produce electricity in a generator.

But for mathematical analysis, of a system all kinds of signals should be represented in a similar form. This is done by transforming all kinds of signal to their Laplace form. Also the transfer function of a system is represented by Laplace form by dividing output Laplace transfer function to input Laplace transfer function. Hence a basic block diagram of a control system can be represented as:

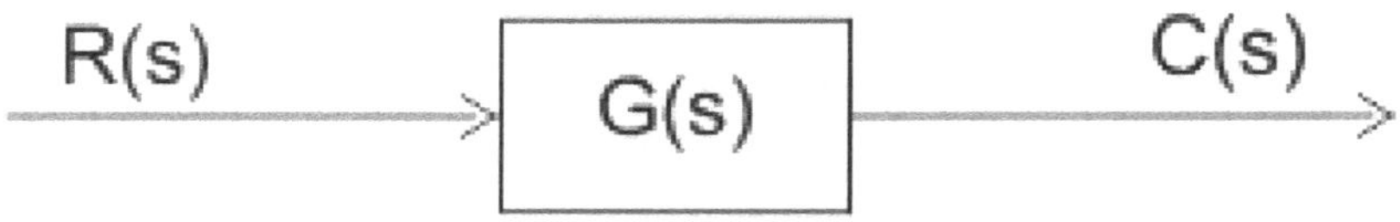

Where,

$$R(s) = £r(t), C(s) = £c(t) \& G(s) = \frac{£c(t)}{£r(t)}$$

Where r(t) and c(t) are time domain function of input and output signal respectively.

Methods of Obtaining a Transfer Function

There are major two ways of obtaining a transfer function for the control system. The ways are:

1. Block Diagram Method: It is not convenient to derive a complete transfer function for a complex control system. Therefore the transfer function of each element of a control system is represented by a block diagram. Block diagram reduction techniques are applied to obtain the desired transfer function.
2. Signal Flow Graphs: The modified form of a block diagram is a signal flow graph. Block diagram gives a pictorial representation of a control system. Signal flow graph further shortens the representation of a control system.

Poles and Zeros of Transfer Function

Generally, a function can be represented to its polynomial form. For example:

$$F(s) = f_0 s^n + f_1 s^{n-1} + f_2 s^{n-2} + f_3 s^{n-3} + \cdots\cdots + f_{n-1} s^1 + f_n$$

Now similarly transfer function of a control system can also be represented as,

$$G(s) = \frac{C(s)}{R(s)} = \frac{C_0 s^n + C_1 s^{n-1} + C_2 s^{n-2} + \cdots\cdots + C_{n-1}s + C_n}{R_0 s^m + R_1 s^{m-1} + R_2 s^{m-2} + \cdots\cdots + R_{m-1}s + R_m}$$

$$= K \frac{(s - z_1)(s - z_2)(s - z_3)\cdots(s - z_n)}{(s - p_1)(s - p_2)(s - p_3)\cdots(s - p_m)}$$

Where K is known as the gain factor of the transfer function.

Now in the above function if $s = z_1$, or $s = z_2$, or $s = z_3, \ldots. s = z_n$, the value of transfer function becomes zero. These z_1, z_2, $z_3, \ldots. z_n$, are roots of the numerator polynomial. As for these roots the numerator polynomial, the transfer function becomes zero, these roots are called zeros of the transfer function.

Now, if $\mathrm{s} = \mathrm{p}_1$, or $\mathrm{s} = \mathrm{p}_2$, or $\mathrm{s} = \mathrm{p}_3, \ldots. \mathrm{s} = \mathrm{p}_\mathrm{m}$, the value of transfer function becomes infinite. Thus the roots of denominator are called the poles of the function.

Now let us rewrite the transfer function in its polynomial form.

$$G(s) = K \frac{(s - z_1)(s - z_2)(s - z_3)\cdots(s - z_n)}{(s - p_1)(s - p_2)(s - p_3)\cdots(s - p_m)}$$

Now, let us consider s approaches to infinity as the roots are all finite number, they can be ignored compared to the infinite s. Therefore,

$$G(s) = K \frac{s^n}{s^m} = K s^{n-m}$$

Hence, when $s \to \infty$ and $n > m$, the function will have also value of infinity, that means the transfer function has poles at infinite s, and the multiplicity or order of such pole is m - n.

Concept of Transfer Function

The transfer function is generally expressed in Laplace Transform and it is nothing but the relation between input and output of a system. Let us consider a system consists of a series connected resistance (R) and inductance (L) across a voltage source (V).

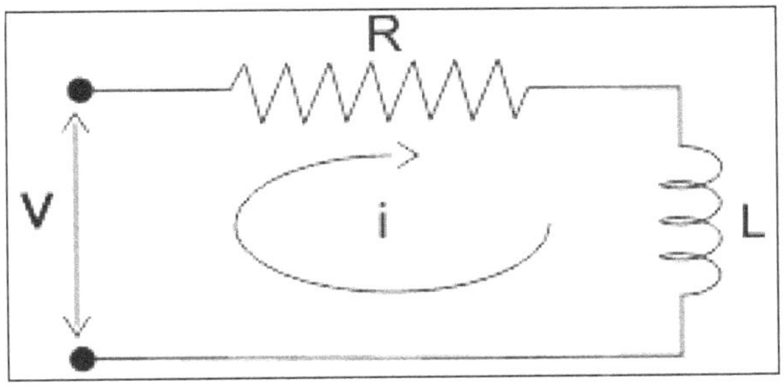

In this circuit, the current 'i' is the response due to applied voltage (V) as cause. Hence the voltage and current of the circuit can be considered as input and output of the system respectively.

From the circuit, we get,

$$V = Ri + L\frac{di}{dt}$$

Now applying Laplace Transform, we get,

$$V(s) = RI(s) + L\left[sI(s) - i(0^+)\right]$$

[$\because$ Initially inductor behaves as open, hence, $i(0^+) = 0$]

$\Rightarrow$ V(s) = I(s)[R + Ls]

$$\Rightarrow \frac{I(s)}{V(s)} = \frac{1}{R + Ls} = \frac{1/L}{s + R/L}$$

The transfer function of the system, $G(s) = I(s)/V(s)$, the ratio of output to input.

1) Let us explain the concept of poles and zeros of transfer function through an example.

$$G(s) = \frac{(s+1)(s+2)}{(s+3)(s+4)(s+5)(s+2-4j)(s+2+4j)}$$

Solution:

The zeros of the function are, -1, -2 and the poles of the functions are $-3, -4, -5, -2+4j, -2-4j$.

Here $n = 2$ and $m = 5$, as $n < m$ and $m - n = 3$, the function will have 3 zeros at $s \to \infty$. The poles and zeros are plotted in the figure below.

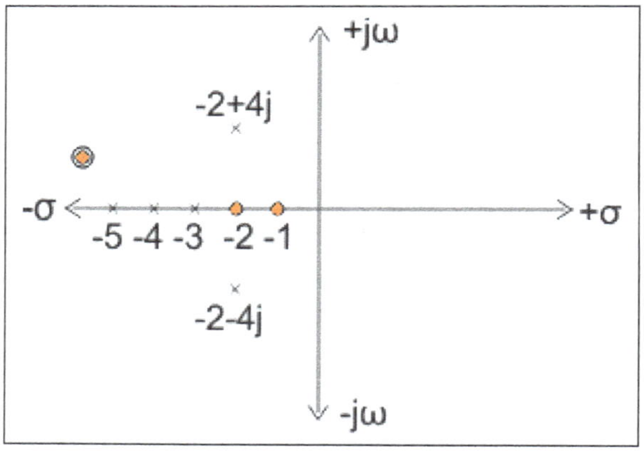

2) Let us take another example of transfer function of control system.

$$G(s) = \frac{(s-2)(s+5)(s+8)}{s(s+1)(s+6)(s+9)(s+1-j3)(s+1+j3)}$$

Solution:

In the above transfer function, if the value of numerator is zero, then

$$(s-2)(s+5)(s+8)=0$$
$$\Rightarrow s=2,-5,-8$$

These are the location of zeros of the function.

Similarly, in the above transfer function, if the value of denominator is zero, then

$$s(s+1)(s+6)(s+9)(s+1-j3)(s+1+j3)=0$$
$$\Rightarrow s=0,-1,-6,-9,-1,+j3,-1-j3$$

These are the location of poles of the function.

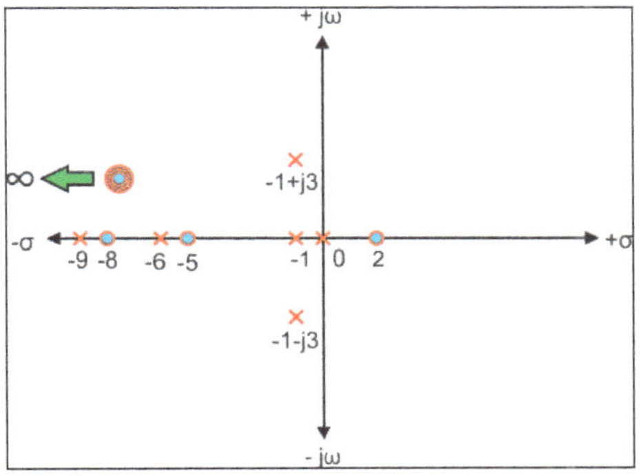

As the number of zeros should be equal to number of poles, the remaining three zeros are located at $s \to \infty$.

Example of transfer function of a network:

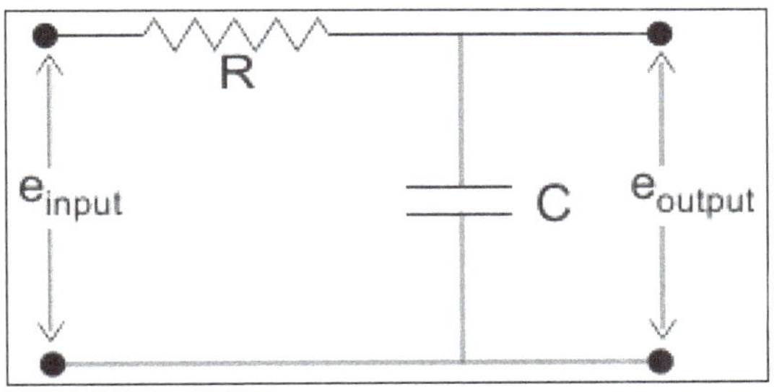

Solution:

In the above network it is obvious that:

$$e_{input} = Ri(t) + \frac{1}{C}\int i(t)dt$$

$$e_{ontput} = \frac{1}{C}\int i(t)dt$$

Let us assume,

$$£\left[e_{input}(t)\right] = E_{input}(s),\ £\left[e_{output}(t)\right] = E_{output}(s),\ £\,[i(t)] = I(s)$$

And then,

$$\pounds\left[\int i(t)dt\right] = \frac{I(s)}{s} + \int i(0)dt = \frac{I(s)}{s} + 0\frac{F(s)}{s}$$

$$\left[\int i(0)dt = 0, \text{ as there is no current initially through the capacitor}\right]$$

Taking the Laplace transform of above equations with considering the initial condition as zero, we get:

$$\frac{E_{output}(s)}{E_{intput}(s)} = \frac{I(s)\dfrac{1}{Cs}}{I(s)\left[R + \dfrac{1}{Cs}\right]} = \frac{\dfrac{1}{Cs}}{\left[R + \dfrac{1}{Cs}\right]} = \frac{1}{RCs + 1}$$

The Effect of Impulse Signal

The unit impulse signal is defined as

$$\delta(\text{t}) = 1 \text{ when, t} = 0$$
$$= 0 \; when, \; t \neq 0$$

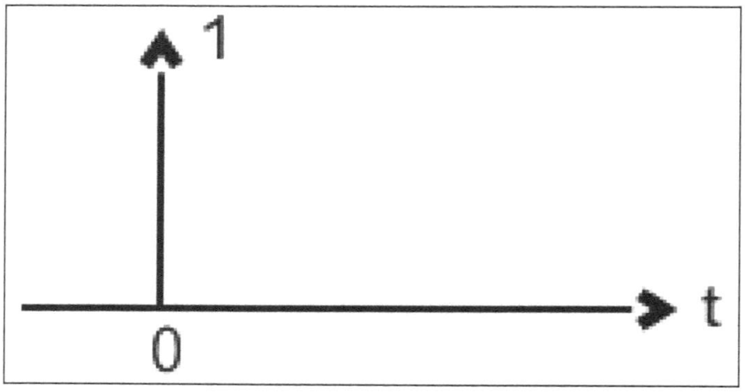

Laplace transform of unit impulse function is 1.

$$\pounds\delta(t) = 1$$

Now if input signal is unit impulse signal then,

$$\frac{C(s)}{R(s)} = G(s) \Rightarrow C(s) = R(s)G(s) \Rightarrow C(s) = G(s)$$

$$[\because \pounds\delta(t) = 1]$$

The output function is same as its transfer function.

Permissions

All chapters in this book are published with permission under the Creative Commons Attribution Share Alike License or equivalent. Every chapter published in this book has been scrutinized by our experts. Their significance has been extensively debated. The topics covered herein carry significant information for a comprehensive understanding. They may even be implemented as practical applications or may be referred to as a beginning point for further studies.

We would like to thank the editorial team for lending their expertise to make the book truly unique. They have played a crucial role in the development of this book. Without their invaluable contributions this book wouldn't have been possible. They have made vital efforts to compile up to date information on the varied aspects of this subject to make this book a valuable addition to the collection of many professionals and students.

This book was conceptualized with the vision of imparting up-to-date and integrated information in this field. To ensure the same, a matchless editorial board was set up. Every individual on the board went through rigorous rounds of assessment to prove their worth. After which they invested a large part of their time researching and compiling the most relevant data for our readers.

The editorial board has been involved in producing this book since its inception. They have spent rigorous hours researching and exploring the diverse topics which have resulted in the successful publishing of this book. They have passed on their knowledge of decades through this book. To expedite this challenging task, the publisher supported the team at every step. A small team of assistant editors was also appointed to further simplify the editing procedure and attain best results for the readers.

Apart from the editorial board, the designing team has also invested a significant amount of their time in understanding the subject and creating the most relevant covers. They scrutinized every image to scout for the most suitable representation of the subject and create an appropriate cover for the book.

The publishing team has been an ardent support to the editorial, designing and production team. Their endless efforts to recruit the best for this project, has resulted in the accomplishment of this book. They are a veteran in the field of academics and their pool of knowledge is as vast as their experience in printing. Their expertise and guidance has proved useful at every step. Their uncompromising quality standards have made this book an exceptional effort. Their encouragement from time to time has been an inspiration for everyone.

The publisher and the editorial board hope that this book will prove to be a valuable piece of knowledge for students, practitioners and scholars across the globe.

Index

www.ingramcontent.com/pod-product-compliance
Lightning Source LLC
LaVergne TN
LVHW080651210926
848485LV00003B/361